Studies in Computational Intelligence 468

Editor-in-Chief

Prof. Janusz Kacprzyk
Systems Research Institute
Polish Academy of Sciences
ul. Newelska 6
01-447 Warsaw
Poland
E-mail: kacprzyk@ibspan.waw.pl

For further volumes:
http://www.springer.com/series/7092

Studies in Computational Intelligence 468

Editor-in-Chief

Agnieszka Dardzinska

Action Rules Mining

 Springer

Author
Dr. Agnieszka Dardzinska
Department of Mechanics and
Applied Computer Science
Bialystok University of Technology
Bialystok
Poland

ISSN 1860-949X   e-ISSN 1860-9503
ISBN 978-3-642-43552-2   ISBN 978-3-642-35650-6 (eBook)
DOI 10.1007/978-3-642-35650-6
Springer Heidelberg New York Dordrecht London

Printed on acid-free paper

Springer is part of Springer Science+Business Media (www.springer.com)

To my great Family

Foreword

Knowledge discovery in databases (KDD, also called data mining) is a well-established branch of informatics. The basis of KDD is created by an analysis of the given data with the goal of finding unsuspected patterns and summarizing the data in new ways to make them understandable and usable for their owners. The analyzed data is often saved in large and complex databases, in some cases covering long history of the corresponding topic. KDD has been for a long time applied in large corporations and in research and development institutions. However, interest is also growing in small and medium-size companies and in the public sector (e.g. *health care*). The growing interest on KDD brings new challenges both for computer scientists and for practitioners. One of the challenges is related to the fact that data mining systems usually can only list results; they are unable to relate the results of mining to the real-world decisions. Thus, a specific challenge is to turn given data into actionable knowledge. The goal is not only to produce strong, yet not known and interesting patterns. The goal is to suggest an action which will be useful for the data owner. This need led to the notion of actionable rules, which was introduced in 1990s. The rule is considered actionable if a data owner can start an action based on this rule and if the action is useful for her/him. Such specification is, however, too wide and makes possible various interpretations. That's why additional definitions were formulated including the definition of an action rule. This was introduced by Ras and Wieczorkowska in 2000. Several branches of research of action rules and similar notions started, and dozens of papers were published. It is notable that important notion of stable and actionable attributes came from the group of researchers led by Professor Zbigniew Ras. The attribute "*Year of birth*" is an example of stable attribute and the attribute "*Level of interest*" is an example of an actionable attribute. These notions represent a clear bridge between analyzed data and relevant items of domain knowledge. There are various theoretically interesting and practically important results on action rules and related topics. It becomes clear that there is a strong need for a comprehensive overview of all achieved results.

Dr. Dardzinska was one of the authors which invented the notion of action rules, and she still continues her research of this topic. She is a person perfectly qualified to prepare such a comprehensive overview. The book gives a smooth introduction to the notion of action rules. Then all necessary concepts are introduced. The core of the book consists in descriptions of many algorithms and methods for mining and dealing with action rules. Attention is given also to handle incomplete values in data. The book emphasizes the theoretical approaches; however, there are enough examples, which make the book well readable. It can be concluded that the book is a carefully written comprehensive description of the state of art in the area of action rules. The book can be seen as a unique source of information on action rules for researchers, teachers and students interested in knowledge discovery in databases.

Prague, October 2012   Jan Rauch

Contents

1

Introduction

Data mining, or knowledge discovery, is frequently referred to in the literature as the process of extracting interesting information or patterns from large databases. There are two major directions in data mining research: patterns and interest. The pattern discovery techniques include: classification, association, and clustering. Interest refers to pattern applications in business, education, medicine, military or other organizations, being useful or meaningful [16]. Since pattern discovery techniques often generate large amounts of knowledge, they require a great deal of expert manual work to post-process the mined results. Therefore, one of the central research problems in the field relates to reducing the volume of the discovered patterns, and selecting appropriate interestingness measures. These measures are intended for selecting and ranking patterns according to their potential interest to the user. Good measures also allow the time and space costs of the mining process to be reduced. There are two aspects of interestingness of rules that have been studied in data mining literature: objective and subjective measures (Liu et al. [26](1997), Adomavicius and Tuzhilin [1](1997) and Silberschatz and Tuzhilin [57](1995)). Objective measures are data-driven and domain-independent. Generally, they evaluate the rules based on the quality and similarity between them. Subjective measures, including unexpectedness, novelty and actionability, are user-driven and domain-dependent. A rule is actionable if a user can do an action to his advantage based on that rule [26]. This definition, in spite of its importance, is quite vague and it leaves an open door to a number of different interpretations of actionability. For instance, we may formulate actions that involve attributes outside the database schema. In order to narrow it down, a new class of rules (called *action rules*) constructed from certain pairs of association rules, extracted from a given database, has been proposed by Ras and Wieczorkowska [52](2000). The main idea was to generate, from a database, special type of rules which basically form a hint to the users, showing a way to re-classify objects with respect to some distinguished attribute (called a *decision attribute*). Values of some of attributes, used to describe objects stored in a database, can be changed and

A. Dardzinska: *Action Rules Mining*, SCI 468, pp. 1–4.
DOI: 10.1007/978-3-642-35650-6_1 © Springer-Verlag Berlin Heidelberg 2013

this change can be influenced and controlled by the user. However, some of
these changes (for instance *profit, treatment*) can not be done directly to a
decision attribute. In addition, the user may be unable or unwilling to proceed with the actions. Independently, another formal definition of an action
rule was proposed in Geffner and Wainer [10](1998). These rules have been
investigated further in [6], [40], [41], [42], [48], [63]. For this reason Tzacheva
and Ras [63](2005) introduced the notion of cost and feasibility of an action
rule. They suggest a heuristic strategy for creating new action rules, where
objects supporting the new action rule also support the initial action rule
but the cost of reclassifying them is lower for the new rule. In this way the
rules constructed are of higher interest to the users. Extended action rules,
discussed in Tsay and Ras [66](2005), and in Dardzinska and Ras [6](2006)
form a special subclass of action rules. They are constructed by extending
headers of action rules in a way that their confidence is increased. The support
of extended action rules is usually lower than the support of the corresponding action rules. A new simplified strategy for action rule extraction was
proposed in Ras and Dardzinska [40](2006), and Ras, Tsay and Dardzinska
[49](2009). In that papers, authors no longer use pairs of classification rules,
but take the objects into consideration. In this sense the action rules are extracted directly from the database. Tzacheva and Ras [64](2007) combine the
approaches of Ras and Dardzinska [40](2006), and Tsay and Ras [66](2005),
to propose an improved constraint based action rule discovery with single
classification rules, which was continued by Ras and Dardzinska [43]. The
minii um support, confidence, and feasibility parameters are specified then
by the user to produce an action rule of desirable low cost [43]. Rauch and
Simunek [47] (2009) in their papers were based on association rules generated by the GUHA method, and were a base to form G-action rules next.
G-action rules can be expressed by two assiociation rules: one for the initial
state (before change), and one for the final state (after change). Yang and
Cheng [71](2002) aim at converting individuals from an undesirable class to
a desirable class. The work proposes actions to move objects to a more desirable class. It is rooted in case-base reasoning, where typical positive cases are
identified to form a small and highly representative case base. This *role model*
is then used to formulate marketing actions. The notion of cost of the action
is also considered. They use 1-NN classifier, 1-cluster-centroid classifier, or
SVM. Such classifiers could become inadequate for disk-resident data due to
their long computational time.

   The work of Ras and associates on action rules is probably pioneering in
the action rule mining [40], [50], [52], [67]. The notions of actionable and
stable attributes can be found from the beginning of their work. In most of
their methods, they use a heuristic rule discovery method first to obtain a
set of rules, then they use a procedure of matching pairs: a rule predicting
the positive class with a related rule predicting the negative class. Unlike
an exhaustive method, their method can miss important rules. Mining action
rules from scratch (He, Xu, Deng, Ma, [16](2005), Ras, Dardzinska, Tsay and

Wasyluk [48](2007), Yang and Cheng [71](2002)), i.e. directly from the database without using pairs of classification rules, or a similar approach, which will present an exhaustive method, would supply us with all important rules. Clearly, the number of set of rules is huge, so a generalization technique, such as creating summaries, provide significant help by reducing the space and furnish the user with the essence of the actionable knowledge. In [48], a new algorithm that discovers action rules directly from a decision system was presented. It gives a completely new approach for generating association-type action rules. The notion of frequent action sets and strategy how to generate them was presented. Authors also introduced the notion of a representative action rule and gave an algorithm to construct them directly from frequent action sets. They presented the notion of a simple association action rule, the cost of association action rule, and gave a strategy to construct simple association action rules of a lowest cost.

To give an example justifying the need of action rules, let us assume that a number of customers have stopped buying products at one of the bakeries. To find the cause of their decision, possibly the smallest and the simplest set of rules describing all these customers is extracted from the customer database.

For instance, let us assume that

$$(Sex, Female) * (Bread, Rye) \rightarrow (Profit, Excellent)$$

is such rule. Assume also that from the same database, a rule

$$(Sex, Female) \rightarrow (Profit, Average)$$

representing the remaining customers has been extracted. At the same time, we know that the bakery stopped baking rye bread some time ago. Now, by comparing these two rules, we can easily find out that the bakery should start selling rye bread again if they do not want to loose more female customers. Action rule, constructed from these two rules, is represented by the expression:

$$(Sex, Female) * (Bread, \rightarrow Rye) \rightarrow (Profit, Average \rightarrow Excellent).$$

It should be read as:

If females will buy rye bread, then they should shift from the average group of customers to the excellent one.

Changing the offer in a bakery is an example of its implementation. A very important measure of interestingness of action rule is its actionability, e.g. whether or not the pattern can be used in the decision making process of a business to increase profit. Hence, recent research focuses on making it easier for the user to grasp the significance of the mined rules, in the context of a business action plan [6], [16], [40], [49], [50], [63]. An action rule provides hints to a business user as to what changes within flexible attributes are needed in order to re-classify customers from low profitability to high profitability class [52]. It is assumed that attributes in a database are divided into two groups: stable and flexible. By stable we mean attributes, whose values cannot be changed (e.g. *age, place of birth, date of birth, gender*). On the other hand, attributes (e.g. *interest rate, loan approval, profit, symptoms, medical treatment*) whose values can be changed or influenced are called flexible.

This book introduces the fundamental concepts necessary for designing, using and implementing action rules. It is designed to be accessible and useful for readers. The goal is to present principal concepts and algorithms for main topic which is action rules mining.

Chapter 1 is a brief introduction to acquaint readers with a new subject. Chapter 2 introduce types of information systems, both complete and incomplete, which were used during the experiments, their terminology and architecture. It also provides methods of finding rules using known algorithms, such as LERS or ERID. In this chapter chase algorithms for handling incomplete values in decision systems are presented. It also discusses distributed information systems, query and transaction processing with data distribution. Chapter 3 covers several topics related to action rules mining. It introduces formal definition of an action rule, methods of extracting action rules from (extracted earlier) classification rules or directly from information systems. It introduces the notion of a simple association action rule, the cost of association action rule, and give a strategy how to construct simple association action rules of a lowest cost. In this chapter a new approach for generating action rules from datasets with numerical attributes by incorporating a tree classifier and a pruning step based on meta-actions is presented. Meta-actions are seen here as a higher-level knowledge about correlations between different attributes.

The algorithms presented in this book were implemented and initially tested in real medical databases, with particular consideration of children flat foot database. Strategies included in this book, comparing to e.g. statistical methods, are innovative in orthopedic area and they seem to be a great chance for creating new approaches. They can be helpful tools for doctors as advisory systems for finding best way of patients'treatment and rehabilitation process. All chapters include many examples corresponding to the material contained in them.

Information Systems

Let us assume that any object in the universe can be described or character-ized by values of some attributes. If descriptions of objects are precise enough with respect to a given concept, we should be able to unambiguously describe a subset of objects representing that concept. However, in many practical situations, the available knowledge is often incomplete and imprecise. This implies, that concepts usually can not be precisely described. Information about objects has to be stored using some data representation model. In this book we use information system proposed in [29] and its table representation as information models about data. Because of the incompleteness of data, information systems may be of different types (different levels of incomplete-ness). Also, the way we interpret the data stored in information system may differ. Finally, since the data are stored, we need tools to retrieve information about them. We can ask for objects satisfying a given description but we can also ask for knowledge which in this book is simplified to rules.

2.1 Types of Information Systems

This section starts with definitions of an information system and a decision system. An information system is used for representing knowledge about a group of objects of a similar type by characterizing them in terms of values of attributes. Its definition, given here, is due to [30]. Assuming that information about objects is often incomplete, and it varies from situation to situation, we present different types of incompleteness of data.

Definition 2.1. *By an Information System we mean a triple $S = (X, A, V)$ where:*

- *X is a nonempty, finite set of objects,*
- *A is a nonempty, finite set of attributes,*
- *$V = \bigcup\{V_a : a \in A\}$ is a set of values of attributes, where V_a is a set of values of attribute a for any $a \in A$.*

A. Dardzinska: *Action Rules Mining*, SCI 468, pp. 5–46.
DOI: 10.1007/978-3-642-35650-6_2 © Springer-Verlag Berlin Heidelberg 2013

We also assume, that:

- $V_a \cap V_b = \emptyset$ *for any* $a, b \in A$ *such that* $a \neq b$
- $a : X \to V_a$ *is a partial function for every* $a \in A$.

Table 2.1 Information System S

X	Attribute a	Attribute b	Attribute c	Attribute d
x_1	a_1	b_1	c_1	d_1
x_2	a_2	b_1	c_1	d_1
x_3	a_3	b_2	c_1	d_2
x_4	a_3	b_2	c_2	d_1
x_5	a_2	b_1	c_2	d_1
x_6	a_2	b_2	c_2	d_2
x_7	a_3	b_1	c_2	d_2
x_8	a_1	b_2	c_3	d_2
x_9	a_1	b_1	c_3	d_2
x_{10}	a_2	b_1	c_3	d_1

Example 2.2. Table 2.1 shows an information system $S = (X, A, V)$, with a set of objects $X = \{x_i\}_{i \in \{1,2,\ldots,10\}}$, set of attributes $A = \{a, b, c, d\}$ and set of their values $V = \{a_1, a_2, a_3, b_1, b_2, c_1, c_2, c_3, d_1, d_2\}$.

If all attributes in S are total functions, then S is a *complete information system*. Otherwise it is called *incomplete information system*.

Table 2.2 Complete Information System

X	Orthopedic_disease	Group_age
x_1	$Haglund - Sever$	*children*
x_2	*Degenerative_arthritis*	*women*
x_3	*Degenerative_arthritis*	*men*

Example 2.3. Information system $S = (X, A, V)$ presented in Table 2.2 is complete, since all its attributes are total functions.

Clearly we can observe, that:
$X = \{x_1, x_2, x_3\}$, $A = \{Orthopedic\_disease, Group\_age\}$, $V = V_O \cup V_G$, where $V_O = V_{Orthopedic\_disease} = \{Haglund-Sever, Degenerative\_arthritis\}$, and $V_G = V_{Group\_age} = \{children, women, men\}$.
$Orthopedic\_disease(x_1) = Haglund - Sever$, $Orthopedic\_disease(x_2) = Orthopedic\_disease(x_3) = Degenerative\_arthritis$,
$Group\_age(x_1) = children$, $Group\_age(x_2) = women$, $Group\_age(x_3) = men$.

Table 2.3 Incomplete Information System

X Orthopedic_disease	Group_age
x_1 Haglund − Sever	children
x_2	women
x_3 Degenerative_arthritis	men

Example 2.4. Information system presented in Table 2.3 is incomplete, since one of the value, *Orthopedic_disease*(x_2), is not defined.

Information systems can be seen as generalizations of decision tables. In any decision table together with the set of attributes a partition of that set into conditions and decisions is given. For simplicity reason, we consider decision tables with only one decision.

As we already have mentioned, there is a number of different types of incompleteness in data. For instance, the value of an attribute for a given object does not have to be known at all, but also it is possible that we may have some incomplete information available about it. The partial information about objects may have many different representations (rough, fuzzy, probabilistic, etc.)[28], [58]. Also, information about data can be incomplete not because it is unknown to the owner of the information system, but e.g. because of the security reasons. Clearly in the latest case, users should not be able to reconstruct any missing data by applying either statistical or rule-discovery based methods.

2.2 Types of Incomplete Information Systems

In this chapter, we present four types of incomplete information systems. A number of possible semantics associated with null values will be covered later in Section 1.3.

Type 1

Let us consider information system $S = (X, A, V)$, as described in Definition 1.1. Incompleteness of the first type is characterized by the assumption that minimum one attribute $a \in A$ is a partial function, which means that

$$(\exists x \in X) \, (\exists a \in A) \, [a(x) = Null]$$

This type is illustrated by Table 1.3. *Null* value is interpreted here as "*undefined*" value.

Type 2.1

For this type of incompleteness, we assume that all attributes in $S = (X, A, V)$ are functions of the type $a : X \to 2^{V_a} - \{\emptyset\}$.

- If $a(x) = \{a_1, a_2, \ldots a_n\} \subseteq V_a$, then we say that *"the value of attribute a is either $a_1, a_2, \ldots a_n$"*.
- If $a(x) = V_a$, then all values of the attribute a are equally probable. For simplicity reason, we represent that fact by $a(x) = Null$. It corresponds to *"blank"* in the table representation of S.

Table 2.4 Information System of Type 2.1

X Name	Surname
x_1 *Mary*	*Smith, Brown*
x_2 *Ann, Mary*	
x_3 *Ann, Betty, Mary*	*Brown*

Example 2.5. Let $S = (X, A, V)$ be an information system of Type 2.1 represented by Table 2.4.

We assume that:

$X = \{x_1, x_2, x_3\}$, $A = \{Name, Surname\}$, $V = N_N \cup V_S$, where
$V_N = V_{Name} = \{Ann, Betty, Mary\}$, $V_S = V_{Surname} = \{Brown, Smith\}$.

The attribute *Name* is defined as:

$Name(x_1) = \{Mary\}$, which is interpreted as *"Name of x_1 is Mary"*.

$Name(x_2) = \{Ann, Mary\}$, which is interpreted as *"Name of x_2 is either Ann or Mary"*.

$Name(x_3) = \{Ann, Betty, Mary\}$, which is interpreted as *"Name of x_3 is either Ann or Betty or Mary"*, and which is equivalent to $Name(x_3) = Null$, as $Name(x_3)$ contains all values from the domain of attribute a.

Type 2.2

For this type of incompleteness, we assume that all attributes in $S = (X, A, V)$ are functions of the type $a : X \to 2^{V_a}$. Type 2.2 is the same as Type 2.1, except that we allow having the empty set as the value of some attributes in S.

If $a(x) = \emptyset$, then we know that the value of attribute a for the object x does not exist.

Table 2.5 Information System of Type 2.2

X Name	Middle_Name	Surname
x_1 *Mary*	*Lucy, Amber*	*Smith, Brown*
x_2 *Mary*	$\emptyset$	*Smith*
x_3	*Lucy*	*Brown*

Example 2.6. Let $S = (X, A, V)$ be an information system of Type 2.2 represented by Table 2.5.

We assume now that
$X = \{x_1, x_2, x_3\}$, $V = V_N \cup V_M \cup V_S$, and the domains of attributes in set of attributes $A = \{Name, Middle\_Name, Surname\}$ are defined as follows:

$V_N = V_{Name} = \{Ann, Betty, Marry\}$,
$V_M = V_{Middle\_Name} = \{Amber, Lucy\}$,
$V_S = V_{Surname} = \{Brown, Smith\}$.

Taking Attribute $Middle\_Name$ into consideration, we have:

$Middle\_Name(x_1) = \{Amber, Lucy\}$, which is interpreted as "*middle name of x_1 is either Amber or Lucy*". It is equivalent to $Middle\_Name(x_1) = Null$, as these values cover the whole domain of the attribute $Middle\_Name$.

$Middle\_Name(x_2) = \emptyset$, which is interpreted as "*object x_2 does not have middle name*". It means that the value is not missing because we know that this value does not exist for x_2.

$Middle\_Name(x_3) = Lucy$, which is interpreted as "*middle name of x_3 is Lucy*".

Type 3

For this type of incompleteness, we assume that all attributes in $S = (X, A, V)$ are functions of the type $a : X \rightarrow 2^{V_a \times R}$. We also assume that if $a(x) = \{(a_1, p_1), (a_2, p_2), \ldots (a_n, p_n)\}$, where p_i is a confidence that object x has property a_i ($i = 1, 2, \ldots, n$), then:

$$\sum_{i=1}^{n} p_i = 1.$$

Table 2.6 Information System of Type 3

X Name	Surname
x_1 $(Mary, 1)$	$Smith, Brown$
x_2 $(Ann, \frac{1}{2}), (Mary, \frac{1}{2})$	$\emptyset$
x_3 $(Ann, \frac{1}{8}), Betty, \frac{3}{8}), (Mary, \frac{1}{2})$	$Brown$

Example 2.7. Let us assume that information system $S = (X, A, V)$ presented in Table 2.6 is of the Type 3, where
$X = \{x_1, x_2, x_3\}$, $A = Name, Surname$, $V = V_N \cup V_S$
and the domains of the attributes $Name, Surname$ are as follows:
$V_N = V_{Name} = \{Ann, Betty, Mary\}$,
$V_S = V_{Surname} = \{Brown, Smith\}$.

For instance, the interpretation of the fact that:

$Name(x_2) = \{(Ann, \frac{1}{2}), (Mary, \frac{1}{2})\}$ will be "*the confidence that x_2 has a name Ann is $\frac{1}{2}$ and that he has a name Mary is $\frac{1}{2}$ as well*".

Also, with respect to the object x_3 we have:

$Name(x_3) = \{Ann\}$ with confidence $\frac{1}{8}$, or $Name(x_3) = \{Betty\}$ with confidence $\frac{3}{8}$, or $Name(x_3) = \{Mary\}$ with confidence $\frac{1}{2}$.

There is another type of attributes called *multi-valued attributes* which we do not investigate here in details, because they can be treated the same way as regular attributes. One of the example of a multi-valued attribute is "*orthopedic disease*". Clearly, many people can have more than one disease which means that the domain of that attribute contains sets instead of single elements. So, sets of values of a multi-valued attribute can be treated as values of that attribute. For instance, if "*orthopedic disease*" is a multi-valued attribute, then $\{rickets\}$ and $\{rickets, flat\_foot\}$ can be treated as its two different attribute values.

2.3   Simple Query Language

To define a query language associated with an information system $S = (X, A, V)$, described in Section 1.2, we have to decide about the elements of the alphabet from which all query expressions are built. In all cases presented so far, the alphabet contains all elements from V and two functors: "$+, *$" [45], [46].

2.3.1   Standard Interpretation of Queries in Complete Information Systems

Definition 2.8. *By a set of simple queries, we mean the smallest set Q satisfying the following conditions:*

- *if $v \in V$, then $v \in Q$,*
- *if $t_1, t_2 \in Q$, then $t_1 + t_2 \in Q$ and $t_1 * t_2 \in Q$.*

Our query language is rather simplified but for the purpose of this section such a restricted version of query languages is sufficient. Its unrestricted version, which clearly contains negation, is not needed here. Queries considered in this section are called simple. Since we already agreed about the syntax of a query language, we need to talk now about its different possible semantics.

Definition 2.9. *Standard interpretation N_S of simple queries in a complete information system $S = (X, A, V)$ is defined as:*

1. *$N_S(v) = \{x \in X : a(x) = v\}$ for any $v \in V_a$,*
2. *$N_S(v_1 + v_2) = N_S(v_1) \cup N_S(v_2)$ and $N_S(v_1 * v_2) = N_S(v_1) \cap N_S(v_2)$.*

Table 2.7 Information System of Type 3

X	Attribute a	Attribute b
x_1	a_1	b_1
x_2	a_2	b_2
x_3	a_2	b_2
x_4	a_1	b_2

Example 2.10. Let us take information system $S = (X, A, V)$, represented by Table 2.7, where each attribute $a \in A$ is a total function. In this example we have:
$X = \{x_1, x_2, x_3, x_4\}$, $A = \{a, b\}$ where $V_a = \{a_1, a_2\}$ and $V_b = \{b_1, b_2\}$.
The alphabet of a simple query language associated with S contains the following symbols: $\{+, *, a_1, a_2, b_1, b_2\}$.

Standard interpretation N_S of a few simple queries built over this alphabet is given below:

$N_S(a_1) = \{x_1, x_4\}$
$N_S(a_1 + b_1) = N_S(a_1) \cup N_S(b_1) = \{x_1, x_4\}$
$N_S(a_2 * b_2) = N_S(a_2) \cap N_S(b_2) = \{x_2, x_3\}$.

2.3.2 Standard Interpretation of Queries in Incomplete Information Systems

In a case when information system is incomplete, two standard interpretations of simple queries (pessimistic and optimistic) can be taken into consideration.

Optimistic Interpretation of Type 1

1. $N_S(v) = \{x \in X : a(x) = v \text{ or } a(x) = Null\}$ for any $v \in V_a$,
2. $N_S(v_1 + v_2) = N_S(v_1) \cup N_S(v_2)$
   $N_S(v_1 * v_2) = N_S(v_1) \cap N_S(v_2)$.

Pessimistic Interpretation of Type 1

1. $N_S(v) = \{x \in X : a(x) = v\}$ for any $v \in V_a$,
2. $N_S(v_1 + v_2) = N_S(v_1) \cup N_S(v_2)$
   $N_S(v_1 * v_2) = N_S(v_1) \cap N_S(v_2)$.

Example 2.11. Let us take incomplete information system represented by Table 2.8.

We assume here that:

$X = \{x_1, x_2, x_3, x_4\}$, $A = \{a, b\}$, $V = V_a \cup V_b$, where $V_a = \{a_1, a_2\}$ and $V_b = \{b_1, b_2\}$.

Table 2.8 Incomplete Information System

X	Attribute a	Attribute b
x_1	a_1	b_1
x_2		b_2
x_3	a_2	
x_4	a_1	b_2

Table 2.9 Two interpretations of the incomplete information system of Type 1

Optimistic Interpretation	Pessimistic Interpretation
$N(a_1) = \{x_1, x_2, x_4\}$	$N(a_1) = \{x_1, x_4\}$
$N(a_2) = \{x_2, x_3\}$	$N(a_2) = \{x_3\}$
$N(b_1) = \{x_1, x_3\}$	$N(b_1) = \{x_1\}$
$N(b_2) = \{x_2, x_3, x_4\}$	$N(b_2) = \{x_2, x_4\}$

Two interpretations forming jointly so called rough-semantics is presented in Table 2.9. Assume that the information system $S = (X, A, V)$ is of the Type 2. The definitions for optimistic and pessimistic interpretations of simple queries are the same for the Type 2.1 and Type 2.2.

Optimistic Interpretation of Type 2.1

1. $N_S(v) = \{x \in X : a(x) = v$ or $v \in a(x)$ or $a(x) = Null\}$ for any $v \in V_a$,
2. $N_S(v_1 + v_2) = N_S(v_1) \cup N_S(v_2)$ and $N_S(v_1 * v_2) = N_S(v_1) \cap N_S(v_2)$.

Pessimistic Interpretation of Type 2.1

1. $N_S(v) = \{x \in X : a(x) = v\}$ for any $v \in V_a$,
2. $N_S(v_1 + v_2) = N_S(v_1) \cup N_S(v_2)$ and $N_S(v_1 * v_2) = N_S(v_1) \cap N_S(v_2)$.

Table 2.10 Incomplete Information System

X	Attribute a	Attribute b
x_1	a_1, a_2	b_1
x_2	a_1	b_1, b_2
x_3		b_1
x_4	a_2, a_3	
x_5	a_1, a_3	b_2

Example 2.12. Assume that a partially incomplete information system $S = (X, A, V)$ is represented by the Table 2.10.

We also assume that:

$X = \{x_1, x_2, x_3, x_4, x_5\}$, $A = \{a, b\}$, $V = V_a \cup V_b$, where $V_a = \{a_1, a_2, a_3\}$ and $V_b = \{b_1, b_2\}$.

Table 2.11 Two interpretations of the incomplete information system of Type 2

Optimistic Interpretation	Pessimistic Interpretation
$N(a_1) = \{x_1, x_2, x_3, x_5\}$	$N(a_1) = \{x_2\}$
$N(a_2) = \{x_1, x_3, x_4\}$	$N(a_2) = \emptyset$
$N(a_3) = \{x_3, x_4, x_5\}$	$N(a_3) = \emptyset$
$N(b_1) = \{x_1, x_2, x_3, x_4\}$	$N(b_1) = \{x_1, x_3\}$
$N(b_2) = \{x_2, x_4, x_5\}$	$N(b_2) = \{x_5\}$

Two interpretations forming jointly so called rough-semantics are given in Table 2.11. Assume that the information system $S = (X, A, V)$ is of the Type 3. Therefore, sets of weighted values of attributes can be used as values of attributes. The interpretation of queries given below was proposed the first time by Ras and Joshi [45]. They also gave its complete and sound set of axioms in [46].

Standard Interpretation of Type 3

1. $N_S(v) = \{(x, p) : (v, p) \in a(x)\}$ for any $v \in V_a$,
2. $N_S(v_1 + v_2) = N_S(v_1) \oplus N_S(v_2)$
3. $N_S(v_1 * v_2) = N_S(v_1) \otimes N_S(v_2)$,

where for any

- $N_S(v_1) = \{(x_i, p_i)\}_{i \in I}$ and $N_S(v_2) = \{(x_j, q_j)\}_{j \in J}$
- $N_S(v_1) \oplus N_S(v_2) =$
  $= \{(x_i, max(p_i, q_i))\}_{i \in I \cap J} \cup \{(x_i, p_i)\}_{i \in I \setminus J} \cup \{(x_j, q_j)\}_{j \in J \setminus I}$
- $N_S(v_1) \otimes N_S(v_2) = \{(x_i, p_i \cdot q_i)\}_{i \in I \cap J}$

Table 2.12 Partially incomplete information system

X	Attribute a	Attribute b
x_1	$(a_1, \frac{1}{2}), (a_2, \frac{1}{2})$	$(b_1, 1)$
x_2	$(a_1, 1)$	$(b_1, \frac{2}{3}), (b_2, \frac{1}{3})$
x_3		$(b_1, \frac{2}{3}), (b_2, \frac{1}{3})$
x_4	$(a_1, \frac{1}{2}), (a_2, \frac{1}{8}), (a_3, \frac{3}{8})$	
x_5	$(a_1, \frac{1}{3}), (a_3, \frac{2}{3})$	$(b_2, 1)$

Example 2.13. Assume that a partially incomplete information system $S = (X, A, V)$ is represented by Table 2.12.

We also assume that:

$X = \{x_1, x_2, x_3, x_4, x_5\}$, $A = \{a, b\}$, $V = V_a \cup V_b$, where $V_a = \{a_1, a_2, a_3\}$ and $V_b = \{b_1, b_2\}$.

Here we have:

$N_S(a_1) = \{(x_1, \frac{1}{2}), (x_2, 1), (x_4, \frac{1}{2}), (x_5, \frac{1}{3})\}$
$N_S(a_2) = \{(x_1, \frac{1}{2}), (x_4, \frac{1}{8})\}$
$N_S(a_3) = \{(x_4, \frac{3}{8}), (x_5, \frac{2}{3})\}$
$N_S(b_1) = \{(x_1, 1), (x_2, \frac{2}{3}), (x_3, \frac{2}{3})\}$
$N_S(b_2) = \{(x_2, \frac{1}{3}), (x_3, \frac{1}{3})\}$

For sample queries of length 2, we get:

$N_S(a_1 + a_2) = N_S(a_1) \oplus N_S(a_2) = \{(x_1, \frac{1}{2}), (x_2, 1), (x_4, \frac{5}{8}), (x_5, \frac{1}{3})\}$
$N_S(a_1 * a_2) = N_S(a_1) \otimes N_S(b_2) = \{(x_2, \frac{1}{3})\}$.

2.4 Rules

Data mining is seen as the efficient discovery of patterns in datasets. Patterns in the data can be represented in many different forms, including units of knowledge called rules. Each rule has a form:

If *set of conditions* **then** *action.*

The left side and also the right side of the rule may involve a single attribute value or a conjunction of attribute values and/or their negations taken from domains of different attributes [13].

Table 2.13 Complete information system S

X	Attribute a	Attribute b	Attribute c	Attribute d
x_1	a_1	b_1	c_2	d_1
x_2	a_2	b_1	c_3	d_2
x_3	a_3	b_2	c_1	d_2
x_4	a_1	b_2	c_2	d_2
x_5	a_2	b_2	c_3	d_2

Example 2.14. Let us take the information system $S = (X, A, V)$ as defined by Table 2.13 and assume that $\{d\}$ is a distinguished attribute called a decision attribute. Attributes $\{a, b, c\}$ are classification attributes in S.

In this case:

$X = \{x_1, x_2, x_3, x_4, x_5\}$ is a set of objects, $A = \{a, b, c, d\}$ is a set of attributes and $V = V_a \cup V_b \cup V_c \cup V_d$ is a set of values of attributes, where $V_a = \{a_1, a_2, a_3\}$, $V_b = \{b_1, b_2\}$, $V_c = \{c_1, c_2, c_3\}$ and $V_d = \{d_1, d_2\}$ are the domains of attributes a, b, c and d respectively.

The decision corresponding to object x_1 in S can be described as:

If the value of attribute a is a_1
    and the value of attribute b is b_1
        and the value of attribute c is c_2
            then the value of attribute d is d_1.

The above rule can be written in an alternate, more formal, way as:

$$[(a(x_1) = a_1) * (b(x_1) = b_1) * (c(x_1) = c_2)] \rightarrow [d(x_1) = d_1]$$

The above rule is true not only for x_1 but also can be true for any object x in S. This is because there is no other object characterized by the same values of a, b and c and characterized by another value for attribute in S. Therefore, this rule can be written as:

$$[(a(x) = a_1) * (b(x) = b_1) * (c(x) = c_2)] \rightarrow [d(x) = d_1]$$

Finally, the rule can be simplified to:

$$[(a(x) = a_1) * (b(x) = b_1)] \rightarrow [d(x) = d_1]$$

as the condition about attribute c is not necessary.

Now, let us consider objects x_3, x_4, x_5 in S. From the Table 2.13 we can directly extract three rules:

$$[(a(x) = a_3) * (b(x) = b_2) * (c(x) = c_1)] \rightarrow [d(x) = d_2]$$
$$[(a(x) = a_1) * (b(x) = b_2) * (c(x) = c_2)] \rightarrow [d(x) = d_2]$$
$$[(a(x) = a_2) * (b(x) = b_2) * (c(x) = c_3)] \rightarrow [d(x) = d_2]$$

These rules can be replaced by one simple rule:

$$[(b(x) = b_2) \rightarrow [d(x) = d_2],$$

as the condition part of the rule does not depend on values of attributes a and c. Finally, simplified rules can be written in even more simplified form given below:

$$a_1 * b_1 \rightarrow d_1, \text{ and } b_2 \rightarrow d_2$$

as they describe the same object x.

Definition 2.15. *By a set of terms for $S = (X, A, V)$ we mean a least set T such that:*

- $0, 1 \in T$,
- $w \in T$, $\sim w \in T$ *for any* $w \in V$,
- *if* $t_1, t_2 \in T$ *then* $(t_1 + t_2) \in T$, $(t_1 * t_2) \in T$.

Definition 2.16. *Term t is called simple, if $t = t_1 * t_2 * \ldots t_n$ and:*

$$(\forall j \in \{1, 2, \ldots n\}) \ [(t_j \in V) \ or \ (t_j =\sim w \wedge w \in V)]$$

Definition 2.17. *Semantics M of terms in $S = (X, A, V)$ is defined in a standard way as:*

- $M(0) = \emptyset$,
- $M(1) = X$,
- $M(w) = \{x \in X : w \in a(x)\}$ *for any* $w \in V_a$,
- $M(\sim w) = X \setminus M(w)$ *for any* $w \in V_a$,
- *if* t_1, t_2 *are terms, then:*
  - $M(t_1 + t_2) = M(t_1) \cup M(t_2)$
  - $M(t_1 * t_2) = M(t_1) \cap M(t_2)$.

Definition 2.18. *By a rule in a decision system $S = (X, A \cup \{d\})$ we mean any structure of the form $t \to d_1$, where t is a simple term for S and $d_1 \in Dom(d)$. If $M(t) \subseteq M(d_1)$, then $t \to d_1$ is called a certain rule.*

Definition 2.19. *Support of the rule $t \to d_1$ (denoted as $sup(t \to d_1)$) is defined as $sup(t * d_1)$ which means the number of objects in S having property $t * d_1$.*

Definition 2.20. *Confidence of a rule $t \to d_1$ (denoted as $conf(t \to d_1)$) is defined as $\frac{sup(t \to d_1)}{sup(t)}$.*

2.5 Distributed Information Systems

In this section we recall the notion of a distributed information system as introduced by Ras and Joshi in [45], [46], and later applied in [35], [36], [37], [38], [39].

Definition 2.21. *By a distributed information system we mean a pair $DS = (\{S_i\}_{i \in I}, L)$ where:*

- $S_i = (X_i, A_i, V_i)$ *is an information system for any $i \in I$, and $V_i = \bigcup \{V_{ia} : a \in A_i\}$,*
- *L is a symmetric, binary relation on the set I,*
- *I is a set of sites.*

Let $S_i = (X_i, A_i, V_i)$ for any $i \in I$. From now on we use A to denote the set of all attributes in DS, $A = \bigcup \{A_i : i \in I\}$. Similarly, $V = \bigcup \{V_i : i \in I\}$.

Definition 2.22. *A distributed information system is object-consistent if the following condition holds:*

$$(\forall i)(\forall j)(\forall x \in X_i \cap X_j)(\forall a \in A_i \cap A_j) \; [(a_{[S_i]}(x) \subseteq a_{[S_j]}(x)) \; or$$
$$(a_{[S_j]}(x) \subseteq a_{[S_i]}(x))],$$

where a_S denotes that a is an attribute in S.

We also assume that $S_j = (X_j, A_j, V_j)$ and $V_j = \bigcup\{V_{ja} : a \in A_j\}$, for any $j \in I$. The inclusion $((a_{[S_i]}(x) \subseteq a_{[S_j]}(x))$ means that the system S_i has more precise information about the attribute a in object x than system S_j. Objects-consistency means that information about objects in one of the systems is either the same or more general than in the other. Saying other words, two consistent systems can not have conflicting information about any object x which is stored in both of them. System in which the above condition does not hold is called *object-inconsistent*.

We assume that any site of *DIS* can be queried either for objects or for knowledge (simplified to rules in this book). A query is built from values of attributes in $V = \bigcup\{V_i : i \in I\}$.

A query is called *local* for site i, if is build only from values in V_i. Otherwise the query is called *global* for site i.

Definition 2.23. *By an incomplete distributed information system we mean a pair $DS = (\{S_i\}_{i \in I}, L)$ where:*

- *$DS = (\{S_i\}_{i \in I}, L)$ is an information system for $i \in I$, where $V_i = \bigcup\{V_{ia} : a \in A_i\}$,*
- *$(\exists i \in I)$ S_i is incomplete,*
- *L is a symmetric, binary relation on the set I,*
- *I is a set of sites.*

Two systems S_i, S_j are called *neighbors in distributed information system* if $(i, j) \in L$.

Let us present a definition of $s(i) - terms$, $s(i) - formulas$ and their standard interpretation M_i in distributed information system $DS = (\{S_j\}_{j \in I}, L)$, where $S_j = (X_j, A_j, V_j)$ and $V_j = \bigcup\{V_{ja} : a \in A_j\}$.

Definition 2.24. *By a set of $s(i) - terms$ for S we mean a least set T_i such that:*

- *$0, 1 \in T_i$,*
- *$(a, w) \in T_i$, $\sim (a, w) \in T_i$ for any $w \in V_{ia}$*
- *if $t_1, t_2 \in T_i$ then $(t_1 + t_2) \in T_i$, $(t_1 * t_2) \in T_i$.*

We say that:

- $s(i)$- *term* is *atomic*, if it has a form (a, w) or $\sim (a, w)$, where $a \in B_i \subseteq A_i$,
- $s(i)$- *term* is *positive*, if it has a form $\prod\{(a, w) : a \in B_i \subseteq A_i$ and $w \in V_{ia}\}$,
- $s(i)$- *term* is *primitive*, if it has a form $\prod\{t_j : t_j$ is *atomic*$\}$,
- $s(i)$- *term* is in DNF, if $t = \sum\{t_j : j \in J\}$ and each t_j is primitive.

If an attribute a is known, than instead of (a, w), we often write just w. Also, instead of (a, v) sometimes we will write v_a.

Definition 2.25. *By a local query for site i ($s(i) - query$) we mean any element in T_i which is in disjunctive normal form (DNF).*

Let us assume that X is a set of objects. By an $X - algebra$ ([45], [46]) we mean a sequence $(N, \oplus, \otimes, \sim)$ where:

- $N = \{N_i : i \in J\}$ where $N_i = \{(x, p_{\langle x,i \rangle}) : p_{\langle x,i \rangle} \in [0, 1] \wedge x \in X\}$
- N is closed under the operations:

- $N_i \oplus N_j = \{(x, max(p_{\langle x,i \rangle}, p_{\langle x,j \rangle})) : x \in X\}$,
- $N_i \otimes N_j = \{(x, p_{\langle x,i \rangle} \cdot p_{\langle x,j \rangle}) : x \in X\}$,
- $\sim N_i = \{(x, 1 - p_{\langle x,i \rangle}\}$.

Theorem: If $N_i, N_j, N_k \in N$, then:

- $N_i \oplus N_i = N_i$,
- $N_i \oplus N_j = N_j \oplus N_i$,
- $N_i \otimes N_j = N_j \otimes N_i$,
- $(N_i \oplus N_j) \oplus N_k = N_i \oplus (N_j \oplus N_k)$,
- $(N_i \otimes N_j) \otimes N_k = N_i \otimes (N_j \otimes N_k)$,
- $(N_i \oplus N_j) \otimes N_k = (N_i \otimes N_k) \oplus (N_j \otimes N_k)$.

Definition 2.26. *A standard interpretation of $s(i) - queries$ in system $DS = (\{S_j\}_{j \in I}, L)$ is a partial function M_i from the set of $s(i) - queries$ into $X_i - algebra$ defined as follows:*

- $dom(M_i) \subseteq T_i$,
- $M_i(a, w) = \{(x, p) : x \in X_i \wedge w \in V_{ia} \wedge p = \frac{1}{card(V_{ia})}\}$,
- $M_i(\sim (a, w)) = \sim M_i(a, w)$,
- *for any atomic term $t_1(a) \in \{(a, w), \sim (a, w)\}$, and any primitive term $t = \prod\{s(b) : [s(b) = (b, w_b)$ or $s(b) = \sim (b, w_b)] \wedge (b \in B_i \subseteq A_i) \wedge (w_b \in V_{ib})\}$ we have:*

$M_i(t * t_1(a)) = M_i(t) \otimes M_i(t_1)$, if $a \notin B_i$
$M_i(t * t_1(a)) = M_i(t)$, if $a \in B_i$ and $t_1(a) = s(a)$
$M_i(t * t_1(a)) = \emptyset$, if $a \in B_i$ and $t_1(a) \neq s(a)$.

- *for any $s(i) - terms$ t_1, t_2 we have:*
  $M_i(t_1 + t_1) = M_i(t_1) \oplus M_i(t_2)$.

Definition 2.27. *By* $(k, i) - rule$ *in* $DS = (\{S_j\}_{j \in I}, L)$ *we mean a pair* $r = (t, c)$ *such that* $c \in V_k - V_i$, t *is a positive* $s(k) - term$ *belonging to the intersection* $T_k \cap T_i$, *and if* $(x, p) \in M_k(t)$ *then* $(x, q) \in M_k(c)$.

So, $(k, i) - rules$ are rules which are discovered at the information system S_k but all their condition attributes belong also to the set of attributes A_i. It means that $(k, i) - rules$ can be seen as definitions of values of attributes which are applicable to both sites k and i. At site k, any $(k, i) - rule$ r can be used to approximate incomplete values of the decision attribute of r. At site i, any $(k, i) - rule$ r can be used to learn a new attribute which is the decision attribute of r.

2.6 Decision Systems

As we mentioned in Section 1.1 a decision system is a special case of information system. We say that an information system $S = (X, A, V)$ is a decision system, if $A = A_{St} \cup A_{Fl} \cup \{d\}$, where $\{d\}$ is a distinguished attribute called the decision. Attributes in A_{St} are called stable while attributes in A_{Fl} are called flexible. They jointly form the set of conditional attributes. *"Date of birth"* or *"gender"* are the examples of stable attributes. *"Medical treatment"* or *"force pressure"* for each patient is an example of flexible attribute. In paper by Ras and Tzacheva [50], a new subclass of attributes called semi-stable attributes is introduced. Semi-stable attributes are typically a function of time, and undergo deterministic changes (for instance attribute *"age"* or *"height"*). In general, we assume that the set of conditional attributes is partitioned into stable, semi-stable, and flexible conditions, but sometimes we divide attributes simply on stable and flexible. Different interpretations, called non-standard, of semi-stable attributes may exist, and in such cases all these attributes can be treated the same way as flexible attributes. In the algorithm of action rule extraction, proposed in [41], [52] attributes which are not flexible are highly undesirable.

Example 2.28. As an example of decision system we take $S = (X, A, V)$, where $X = \{x_i\}_{i \in \{1,2,...,10\}}$ is a set of objects, $A = \{a, c\} \cup \{b\} \cup \{d\}$ a set of attributes with $A_{St} = \{a, c\}$ - stable attributes, $A_{Fl} = \{b\}$ - flexible attribute and the decision attribute $\{d\}$. Table 2.14 describes a part of medical database, where attributes a, c denote gender and age of a patient, attribute b describes types of medicines, d is a kind of treatment of a patient (H denotes hospital treatment, while A denotes outpatient treatment).

2.7 Partially Incomplete Information Systems

The decision system described in previous section is generally a complete information system, but we may also consider the possibility of having

Table 2.14 Decision System S

X	Attribute a	Attribute b	Attribute c	Decision d
x_1	a_1	b_1	c_1	H
x_2	a_2	b_2	c_3	H
x_3	a_2	b_1	c_3	A
x_4	a_3	b_1	c_2	A
x_5	a_2	b_1	c_2	A
x_6	a_2	b_2	c_2	H
x_7	a_3	b_1	c_2	A
x_8	a_1	b_2	c_1	A
x_9	a_1	b_1	c_3	H
x_{10}	a_2	b_2	c_3	H

incomplete information system. There is a number of different types of incompleteness in data. For instance, the value of an attribute for a given object does not have to be known at all, but also it is possible that we may have some incomplete information available about it. The partial information about objects may have many different representations. Also, information about data can be incomplete not because it is unknown to the user of the information system, but e.g. the user does not want to express his knowledge to other people. Clearly in the latest case, users should not be able to reconstruct any missing data by applying either statistical or rule-discovery based methods. For categorical attributes with low cardinality domains rule induction techniques such as decision tree and decision systems can be used to derive the missing values. However, for categorical attributes with large cardinality domains, the rule induction techniques may suffer due to too many predicted classes. The discovered association relationships among different attributes can be thought as constraint information of their possible values and can be used to predict the true values of missing attributes [4]. So, the assumption placed on incompleteness of data allows having a set of weighted attribute values as a value of an attribute [7]. Additionally, we assume that the sum of these weights has to be equal 1.

Definition 2.29. *We say that $S = (X, A, V)$ is partially incomplete information system of type λ, if S is an incomplete information system and the following three conditions hold:*

- $a_S(x)$ *is defined for any $x \in X$, $a \in A$*
- $\forall (x \in X) \forall (a \in A)[(a_S(x) = \{(a_i, p_i) : 1 \leq i \leq m\}) \rightarrow \sum p_i = 1]$
- $\forall (x \in X) \forall (a \in A)[(a_S(x) = \{(a_i, p_i) : 1 \leq i \leq m\}) \rightarrow (\forall i)(p_i \geq \lambda)]$

2.8 Extracting Classification Rules

In order to induce rules in which the **if** part consists of attributes belonging to $A_{St} \cup A_{Fl}$, and the **then** part consists of the decision attribute $\{d\}$, for instance system LERS [13] or ERID [4], [5] can be used for rules extraction. We can extract the set of classification rules from complete or incomplete decision systems as well.

2.8.1 Attribute Dependency and Coverings

Let B and C be two nonempty sets of the set of all attributes A in $S = (X, A, V)$. Set C depends on set B, $(B \rightarrow C)$, if and only if $\widetilde{B} \subseteq \widetilde{C}$ [14].
Note that $B \rightarrow C$ if and only if the attribute dependency inequality $B^* \leq C^*$ holds. It means, that for each equivalence class E of B^* there exist equivalence class E' of C^* such that $E \subseteq E'$. Therefore, if a pair of objects cannot be distinguished by means of elements from B, they cannot also be distinguished by elements from C.

It is obvious, that if C depends on B and B depends on C, then B is equivalent to C, $B^* = C^*$.

Definition 2.30. *Let C and D be two nonempty subsets of the set A. A subset B of the set D is called a covering of C in D if and only if C depends on B and B is minimal in D.*

It can be also interpreted as: a subset B of D is a covering of C in D if and only if C depends on B and no proper subset B' of B exists such that C depends on B'.

The concept of a covering induces rules depending on as small number of attributes as possible. Coverings of C in D should be used, to generate rules in which the antecedent may consist of attributes from at most a set D of attributes and the consequent consists of decisions from a set C.

Therefore, to find a covering B of C in D three conditions must occur:

1. B_{min}, set B must be minimal,
2. $B \subseteq D$, set B must be a subset of set D,
3. $B \rightarrow C$, set C depends on set B.

The last condition is true if and only if one of the following equivalent conditions is true:

3.1 $\widetilde{B} \subseteq \widetilde{C}$

3.2 $B^* \leq C^*$

To check condition 3.1 for each set B, a new indiscernibility relation associated with B must be computed. It can be presented as $\widetilde{B} = \bigcap_{b \in B} \widetilde{\{b\}}$.
In case of condition 3.2 for each set B, a new partition generated by B must be determined. The partition of X generated by B may be expressed as $B^* = \prod_{b \in B} \widetilde{\{b\}}^*$.

2.8.2   System LERS

The learning system *LERS* (*Learning from Examples based on Rough Sets*) induces a set of rules from examples (any kind of data) and classifies new examples using the set of rules induced previously by LERS [13], [14]. These rules are more general than information contained in the original input data since more new examples may be correctly classified by rules than may be matched with examples from the original data. This handles inconsistencies using rough set theory. The main advantage of rough set theory, introduced by Pawlak [30] is, that it does not need any additional information about data (like probability in probability theory, grade of membership in fuzzy set theory, etc.). In rough set theory approach inconsistencies are not removed from consideration [30], [58]. Instead, lower and upper approximations of the concept are computed. On the basis of these approximations, LERS computes two corresponding sets of rules: certain and possible. One of potential applications is the use of expert systems, equipped with rules induced by LERS, as advisory systems, helping in decision making and improvement strategy [13]. Its input data is represented as a decision table. Examples are described by values of attributes and characterized by a value of a decision. All examples with the same value of the decision belong to the same concept. This system looks for some regularities in the decision table.

System LERS has two main options of rule induction, which are a basic algorithm, invoked by selecting Induce Rules from the menu Induce Rule Set (LEM2) (working on the level of attribute-value pairs), and Induce Rules Using Priorities on Concept Level of the menu Induce Rule Set (LEM1)(working on entire attributes).

2.8.3   Algorithm for Finding the Set of All Coverings (LEM1)

Let $S = (X, A, V)$. The aim of algorithm presented below is to find the set K of all coverings of C in D, where $C, D \subset A$. The cardinality of the set X is denoted as $card(X)$. The set of all subsets of the same cardinality k of the set D is denoted as $\mathbf{B}_k = \{\{x_{i_1}, x_{i_2}, ..., x_{i_k}\} : (\forall j \leq k)[x_{i_j} \in D]$, where k is a positive integer.

Algorithm Cover S, C, D
BEGIN

$K := 0$
for each attribute a in D **do**
compute partition $\{a\}^*$ and compute partition C^*
$k := 1$;
while $k \leq card(D)$ **do begin**:
for each set B in $\mathbf{B}_k$ **do**
if (B is not a superset of any member of K) and ($\prod_{x \in B}\{a\}^* \leq C^*$)

then add B to K;
$k := k + 1$
end
END

Below, there is an example showing how to find covering using $LEM1$ module
of $LERS$.

Table 2.15 Complete Information System S

X	Attribute a	Attribute b	Attribute c	Attribute d
x_1	a_1	b_1	c_1	d_1
x_2	a_1	b_2	c_2	d_2
x_3	a_1	b_1	c_1	d_1
x_4	a_1	b_2	c_2	d_2
x_5	a_2	b_2	c_1	d_3
x_6	a_2	b_2	c_1	d_3

Example 2.31. Let $S = (X, A, V)$ be an information system which is represented by Table 2.15. Assume, that $X = \{x_1, x_2, x_3, x_4, x_5, x_6\}$. Let us
also assume that $\{a, b, c\}$ are classification attributes and $\{d\}$ is a decision
attribute.

The partitions of X generated by single attributes are:
$\{a\}^* = \{\{x_1, x_2, x_3, x_4\}, \{x_5, x_6\}\}$
$\{b\}^* = \{\{x_1, x_3\}, \{x_2, x_4, x_5, x_6\}\}$
$\{c\}^* = \{\{x_1, x_3, x_5, x_6\}, \{x_2, x_4\}\}$
For decision attribute $\{d\}$ we have:
$\{d\}^* = \{\{x_1, x_3\}, \{x_2, x_4\}, \{x_5, x_6\}\}$.
Clearly, none of the above is a subset of $\{d\}^*$ (not marked), so we go to next
step, which is forming two item sets.
$\{a, b\}^* = \{\{x_1, x_3\}, \{x_2, x_4\}, \{x_5, x_6\}\} \subseteq \{d\}^*$- **marked**
$\{a, c\}^* = \{\{x_1, x_3\}, \{x_2, x_4\}, \{x_5, x_6\}\} \subseteq \{d\}^*$- **marked**
$\{b, c\}^* = \{\{x_1, x_3\}, \{x_2, x_4\}, \{x_5, x_6\}\} \subseteq \{d\}^*$- **marked**
All of the above sets are marked. So we can't go further. Therefore the
coverings of $C = \{d\}$ in $A = \{a, b, c\}$ are: $\{a, b\}$, $\{a, c\}$ and $\{b, c\}$.

Now, we can proceed to the next step which is extracting rules from coverings. Let us first consider the covering $\{a, b\}$ computed in the previous step.
From this covering we obtain:
$(a, a_1)^* = \{x_1, x_2, x_3, x_4\}$
$(a, a_2)^* = \{x_5, x_6\} \subseteq \{(d, d_3)\}^*$- **marked**
$(b, b_1)^* = \{x_1, x_3\} \subseteq \{(d, d_1)\}^*$- **marked**
$(b, b_2)^* = \{x_2, x_4, x_5, x_6\}$
Remaining (not marked) sets are $(a, a_1)^*$ and $(b, b_2)^*$, so next step is to con-
catenate them. Then we obtain next set:

$((a, a_1), (b, b_2))^* = \{x_2, x_4\} \subseteq \{(d, d_2)\}^*$- **marked**

Because the last set in covering $\{a, b\}$ was marked, the algorithm stopped.

Therefore, the certain rules, obtained from marked items, are as follows:

$(a, a_2) \rightarrow (d, d_3)$

$(b, b_1) \rightarrow (d, d_1)$

$(a, a_1) * (b, b_2) \rightarrow (d, d_2)$.

Possible rules, which come from non-marked items are:

$(a, a_1) \rightarrow (d, d_1)$ with confidence $\frac{1}{2}$

$(a, a_1) \rightarrow (d, d_2)$ with confidence $\frac{1}{2}$

$(b, b_2) \rightarrow (d, d_2)$ with confidence $\frac{1}{2}$

$(b, b_2) \rightarrow (d, d_3)$ with confidence $\frac{1}{2}$.

For covering $\{a, c\}$ we obtain:

$(a, a_1)^* = \{x_1, x_2, x_3, x_4\}$

$(a, a_2)^* = \{x_5, x_6\} \subseteq \{(d, d_3)\}^*$- **marked**

$(c, c_1)^* = \{x_1, x_3, x_5, x_6\}$

$(c, c_2)^* = \{x_2, x_4\} \subseteq \{(d, d_2)\}^*$ - **marked**.

Remaining sets are $(a, a_1)^*$ and $(c, c_1)^*$, so next step is to make a pair from them. Then we obtain next set:

$((a, a_1), (c, c_1))^* = \{x_1, x_3\} \subseteq \{(d, d_1)\}^*$- **marked**

Because the last set in covering $\{a, c\}$ was marked, the algorithm stopped.

The certain rules, obtained from marked items, are as follows:

$(a, a_2) \rightarrow (d, d_3)$

$(c, c_2) \rightarrow (d, d_2)$

$(a, a_1) * (c, c_1) \rightarrow (d, d_1)$.

Possible rules, which come from non-marked items are:

$(a, a_1) \rightarrow (d, d_1)$ with confidence $\frac{1}{2}$

$(a, a_1) \rightarrow (d, d_2)$ with confidence $\frac{1}{2}$

$(c, c_1) \rightarrow (d, d_1)$ with confidence $\frac{1}{2}$

$(c, c_1) \rightarrow (d, d_3)$ with confidence $\frac{1}{2}$.

For covering $\{b, c\}$ we obtain:

$(b, b_1)^* = \{x_1, x_3\} \subseteq \{(d, d_1)\}^*$- **marked**

$(b, b_2)^* = \{x_2, x_4, x_5, x_6\}$

$(c, c_1)^* = \{x_1, x_3, x_5, x_6\}$

$(c, c_2)^* = \{x_2, x_4\} \subseteq \{(d, d_2)\}^*$ - **marked**

Remaining sets are $(b, b_2)^*$ and $(c, c_1)^*$, so next step is to connect them. Then we obtain next set:

$((b, b_2), (c, c_1))^* = \{x_5, x_6\} \subseteq \{(d, d_3)\}^*$- **marked**

Because the last set in covering $\{b, c\}$ was marked, the algorithm stopped.

The certain rules, obtained from marked items, are as follows:

$(b, b_1) \rightarrow (d, d_1)$

$(c, c_2) \rightarrow (d, d_2)$

$(b, b_2) * (c, c_1) \rightarrow (d, d_3)$

Possible rules, which come from non-marked items are:

$(b, b_2) \rightarrow (d, d_2)$ with confidence $\frac{1}{2}$

$(a, a_1) \to (d, d_3)$ with confidence $\frac{1}{2}$
$(c, c_1) \to (d, d_1)$ with confidence $\frac{1}{2}$
$(c, c_1) \to (d, d_3)$ with confidence $\frac{1}{2}$.

Table 2.16 Incomplete Information System S

X	Attribute a	Attribute b	Attribute c	Attribute d
x_1	a_1	b_2	c_1	d_1
x_2	a_3	b_1		d_1
x_3	a_1	b_2	c_2	d_2
x_4	a_1		c_1	d_2
x_5	a_2	b_1	c_1	d_1
x_6	a_3	b_2	c_2	d_2

2.8.4 Algorithm LEM2

Now, let us present our own algorithm for the second module ($LEM2$) of algorithm $LERS$ which finds certain and possible rules for a given value of a decision attribute. We assume here that user provides the threshold value for minimal confidence of rules to be extracted from system S.

Algorithm $LEM2(S, \lambda)$

- S - incomplete decision system
- λ - minimum confidence threshold

BEGIN
$i := 1$
while $i < I$ **do begin**
$j := 1; m := 1;$
while $j < J[i]$ **do begin**
if $conf(a[i, j] \to d(m)) \geq \lambda$ **then begin**
$mark[(a[i, j]^* \subseteq d(m)^*)] = positive;$
$output[(a[i, j] \to d(m);$ **end**
$j := j + 1;$ **end; end**
$I_k := \{i_k\};$ - i_k-*index randomly chosen from* $\{1, 2, ..., I\}$
for all $j_k \leq J(i_k)$ **do** $a[(i_k, j_k)]_{i_k \in I_k} := a(i_k, j_k);$
for all i, j **such that** both rules $a[(i_k, j_k)]_{i_k \in I_k} \to d(m)$ and $a[i, j] \to d(m)$ are not marked and $i \notin I_k$ **do begin**
if $conf(a[(i_k, j_k)]_{i_k \in I_k} \cdot a[i, j] \to d(m)) \geq \lambda;$**then begin**
$mark(a[(i_k, j_k)]_{i_k \in I_k} \cdot a[i, j]^* \subseteq d(m^*)) = positive;$
$output(a[(i_k, j_k)]_{i_k \in I_k} \cdot a[i, j] \to d(m));$ **end**
else begin
$I_k := I_k \cup \{i\};$
$a[(i_k, j_k)]_{i_k \in I_k} := a[(i_k, j_k)]_{i_k \in I_k} \cdot a[i, j];$ **end; end**
END

The main reason to provide a new version for $LEM2$ was to have an algorithm which can be naturally extended to incomplete information systems.

Example 2.32. Let us assume we have incomplete information system $S = (X, A, V)$, where $\{a, b, c\}$ are classification attributes, and $\{d\}$ is a decision attribute, as in Table 2.16. For simplicity, for all attributes, we will use notation a_i instead of (a, a_i).

Using LERS-LEM2 algorithm and a minimum confidence threshold: $\lambda = \frac{3}{4}$ first we create classes related to decision attribute $\{d\}$:

$d_1^* = \{x_1, x_2, x_5\}$
$d_2^* = \{x_3, x_4, x_6\}$

and classes connected with classification attributes.

First loop:

$a_1^* = \{x_1, x_3, x_4\}$	$a_1^* \nsubseteq d_1^*$	(not marked)
	$a_1^* \nsubseteq d_2^*$	(not marked)
$a_2^* = \{x_5\}$	$a_2^* \subseteq d_1^*$	***marked***
$a_3^* = \{x_2, x_6\}$	$a_3^* \nsubseteq d_1^*$	(not marked)
	$a_3^* \nsubseteq d_2^*$	(not marked)
$b_1^* = \{x_2, x_5\}$	$b_1^* \subseteq d_1^*$	***marked***
$b_2^* = \{x_1, x_3, x_6\}$	$b_2^* \nsubseteq d_1^*$	(not marked)
	$b_2^* \nsubseteq d_2^*$	(not marked)
$c_1^* = \{x_1, x_4, x_5\}$	$c_1^* \nsubseteq d_1^*$	(not marked)
	$c_1^* \nsubseteq d_2^*$	(not marked)
$c_2^* = \{x_3, x_6\}$	$c_2^* \subseteq d_2^*$	***marked***

Second loop: (building terms of length 2 from terms that have not been marked):

$(a_1, b_2)^* = \{x_1, x_3\}$	$(a_1, b_2)^* \nsubseteq d_1^*$	(not marked)
	$(a_1, b_2)^* \nsubseteq d_2^*$	(not marked)
$(a_1, c_1)^* = \{x_1, x_4\}$	$(a_1, c_1)^* \nsubseteq d_1^*$	(not marked)
	$(a_1, c_1)^* \nsubseteq d_2^*$	(not marked)
$(a_3, b_2)^* = \{x_6\}$	$(a_3, b_2)^* \subseteq d_2^*$	***marked***
$(a_3, c_1)^* = \emptyset$		***marked***, but no rule
$(b_2, c_1)^* = \{x_1\}$	$(b_2, c_1)^* \subseteq d_1^*$	***marked***

Third loop: (building terms of length 3 from terms of length 2 and length 1 that have not been marked):

The set $(a_1, b_2, c_1)^*$ is not considered as superset of $(b_2, c_1)^*$ which was previously marked.

Now, from all not marked sets, possible rules satisfying the required threshold are produced. So now we check:

$conf[(a_1, b_2)^* \subseteq d_1^*] = \frac{1}{2} < \lambda$ (do not have a rule)
$conf[(a_1, b_2)^* \subseteq d_2^*] = \frac{1}{2} < \lambda$ (do not have a rule)
$conf[(a_1, c_1)^* \subseteq d_1^*] = \frac{1}{2} < \lambda$ (do not have a rule)
$conf[(a_1, c_1)^* \subseteq d_2^*] = \frac{1}{2} < \lambda$ (do not have a rule)
$conf[(a_1^* \subseteq d_1^*] = \frac{1}{3} < \lambda$ (do not have a rule)
$conf[(a_1^* \subseteq d_2^*] = \frac{2}{3} < \lambda$ (do not have a rule)
$conf[(a_3^* \subseteq d_1^*] = \frac{1}{2} < \lambda$ (do not have a rule)
$conf[(a_3^* \subseteq d_2^*] = \frac{1}{2} < \lambda$ (do not have a rule)
$conf[(b_2^* \subseteq d_1^*] = \frac{1}{3} < \lambda$ (do not have a rule)
$conf[(b_2^* \subseteq d_2^*] = \frac{2}{3} < \lambda$ (do not have a rule)
$conf[(c_1^* \subseteq d_1^*] = \frac{2}{3} < \lambda$ (do not have a rule)
$conf[(c_1^* \subseteq d_2^*] = \frac{1}{3} < \lambda$ (do not have a rule)

Rules extracted during the whole process are:

From the first loop:	$a_2 \rightarrow d_1$	confidence 1
	$b_1 \rightarrow d_1$	confidence 1
	$c_2 \rightarrow d_2$	confidence 1
From the second loop:	$a_3 * b_2 \rightarrow d_2$	confidence 1
	$b_2 * c_1 \rightarrow d_1$	confidence 1.

2.8.5 Algorithm for Extracting Rules from Incomplete Decision System (ERID)

The method of Extracting Rules from Incomplete Decision System (simplified to LERS-LEM2 like algorithm) was presented in detail in [4], [5].

Algorithm $ERID(S, \lambda_1, \lambda_2)$

- S - incomplete decision system
- λ_1 - minimum support threshold
- λ_2 - minimum confidence threshold

BEGIN
$i := 1$
while $i < I$ **do begin**
$j := 1;\ m := 1;$
while $j < J[i]$ **do begin**
if $sup(a[i, j] \rightarrow d(m)) < \lambda_1$ **then** $mark[(a[i, j]^* \subseteq d(m)^*)] = negative;$
if $sup(a[i, j] \rightarrow d(m)) \geq \lambda_1$ and $conf(a[i, j] \rightarrow d(m)) \geq \lambda_2)$ **then begin**
$mark[(a[i, j]^* \subseteq d(m)^*)] = positive;$
$output[(a[i, j] \rightarrow d(m);$ **end**
$j := j + 1;$
end; end
$I_k := \{i_k\};$ - i_k-index randomly chosen from $\{1, 2, ..., I\}$
for all $j_k \leq J(i_k)$ **do** $a[(i_k, j_k)]_{i_k \in I_k} := a(i_k, j_k);$
for all i, j **such that** both rules $a[(i_k, j_k)]_{i_k \in I_k} \rightarrow d(m);\ a[i, j] \rightarrow d(m)$ are

not marked and $i \notin I_k$ **do begin**
if $sup(a[(i_k, j_k)]_{i_k \in I_k} \cdot a[i,j] \rightarrow d(m)) < \lambda_1$ **then**
$mark(a[(i_k, j_k)]_{i_k \in I_k} \cdot a[i,j]^* \subseteq d(m^*)) = negative;$
if $sup(a[(i_k, j_k)]_{i_k \in I_k} \cdot a[i,j] \rightarrow d(m)) \geq \lambda_1$ and $conf(a[(i_k, j_k)]_{i_k \in I_k} \cdot a[i,j] \rightarrow$
$d(m)) \geq \lambda_2$ **then begin**
$mark(a[(i_k, j_k)]_{i_k \in I_k} \cdot a[i,j]^* \subseteq d(m^*)) = positive;$
$output(a[(i_k, j_k)]_{i_k \in I_k} \cdot a[i,j] \rightarrow d(m));$ **end**
else begin
$I_k := I_k \cup \{i\};$
$a[(i_k, j_k)]_{i_k \in I_k} := a[(i_k, j_k)]_{i_k \in I_k} \cdot a[i,j];$ **end; end**
END

The complexity of this algorithm is similar to complexity of $LEM2$.

Example 2.33. Let us assume the same decision system as in the previous example (Table 2.16) with minimum confidence threshold: $\lambda = \frac{3}{4}$. Classes related to decision attribute $\{d\}$:

$d_1^* = \{x_1, x_2, x_5\}$ and $d_2^* = \{x_3, x_4, x_6\}$

and classes connected with classification attributes.

First loop:

$a_1^* = \{x_1, x_3, x_4\}$	$conf(a_1^* \subseteq d_1^*) = \frac{1}{3} < \lambda$	(not marked)
	$conf(a_1^* \subseteq d_2^*) = \frac{2}{3} < \lambda$	(not marked)
$a_2^* = \{x_5\}$	$a_2^* \subseteq d_1^*$	*marked*
$a_3^* = \{x_2, x_6\}$	$conf(a_3^* \subseteq d_1^*) = \frac{1}{2} < \lambda$	(not marked)
	$conf(a_3^* \subseteq d_2^*) = \frac{1}{2} < \lambda$	(not marked)
$b_1^* = \{x_2, x_5\}$	$b_1^* \subseteq d_1^*$	*marked*
$b_2^* = \{x_1, x_3, x_6\}$	$conf(b_2^* \subseteq d_1^*) = \frac{1}{3} < \lambda$	(not marked)
	$conf(b_2^* \subseteq d_2^*) = \frac{2}{3} < \lambda$	(not marked)
$c_1^* = \{x_1, x_4, x_5\}$	$conf(c_1^* \subseteq d_1^*) = \frac{2}{3} < \lambda$	(not marked)
	$conf(c_1^* \subseteq d_2^*) = \frac{1}{3} > \lambda$	(not marked)
$c_2^* = \{x_3, x_6\}$	$c_2^* \subseteq d_2^*$	*marked*

Second loop: (building terms of length 2 from terms that have not been marked):

$(a_1, b_2)^* = \{x_1, x_3\}$	$conf[(a_1, b_2)^* \subseteq d_1^*] = \frac{1}{2} < \lambda$	(not marked)
	$conf[(a_1, b_2)^* \subseteq d_2^*] = \frac{1}{2} < \lambda$	(not marked)
$(a_1, c_1)^* = \{x_1, x_4\}$	$conf[(a_1, c_1)^* \subseteq d_1^*] = \frac{1}{2} < \lambda$	(not marked)
	$conf[(a_1, c_1)^* \subseteq d_2^*] = \frac{1}{2} < \lambda$	(not marked)
$(a_3, b_2)^* = \{x_6\}$	$(a_3, b_2)^* \subseteq d_2^*$	*marked*
$(a_3, c_1)^* = \emptyset$		*marked*, but no rule
$(b_2, c_1)^* = \{x_1\}$	$(b_2, c_1)^* \subseteq d_1^*$	*marked*

Third loop: (building terms of length 3 from terms of length 2 and length 1 that have not been marked):

The set $(a_1, b_2, c_1)^*$ is not considered as superset of $(b_2, c_1)^*$ which was marked.

Rules extracted during the whole process are:

From the first loop:	$a_2 \to d_1$	confidence 1, support 1
	$b_1 \to d_1$	confidence 1, support 1
	$c_2 \to d_2$	confidence 1, support 2
From the second loop:	$a_3 * b_2 \to d_2$	confidence 1, support 1
	$b_2 * c_1 \to d_1$	confidence 1, support 1.

2.9 Chase Algorithms

Common problems either for Information Systems (IS) or for Distributed Information Systems (DIS) include handling of incomplete attributes. One of the solution involves the generation of rules describing all incomplete attributes and then *chasing* the unknown values in the local database with respect to the generated rules. These rules can be given by domain experts but also can be discovered locally or at other sites of DIS. Since all unknown values would not necessarily be found, the process is repeated on the enhanced database until all unknowns are found or no new information can be generated. Before we cover rule-based chase algorithms, chase algorithm based on functional dependencies will be presented [4].

2.9.1 Tableaux Systems and Chase

In this section we introduce a few notions that are fundamental in proving results. The basic notion is that of *tableaux system* , which is defined as an information system, except that it may contain other symbols, called *variables*, beside attribute values from the domains [8]. The notion of *containment mapping* will be introduced to formalize this representation of information systems by means of tableaux systems [11].

Table 2.17 Tableaux System

X	Nurse_Name	Hospital_Ward	Head
x_1	Mary	v_d	n_1
x_2	Ann	v_d	Jones
x_3	Betty	n_2	n_3
x_4	v_E	surgical	n_4
x_5	Lucy	surgical	n_5

Table 2.17 shows a tableaux system while Table 2.18 shows an information system obtained from previous table by replacing variables with constants.

The final notion, the *chase*, is based on the following observation. If the information system represented by a tableaux system is known to satisfy some given constraints, then its variables cannot freely represent contants.

Table 2.18 Information System obtained from Table 2.17

X	Nurse_Name	Hospital_Ward	Head
x_1	Mary	pediatric	Jones
x_2	Ann	pediatric	Jones
x_3	Betty	maternity	Smith
x_4	Kate	surgical	Brown
x_5	Lucy	surgical	Brown

For example, given the tableaux system in Table 2.17, if the associated relations satisfy functional dependency $Hospital\_Ward \rightarrow Head$, then we can infer that variable n_1 must correspond to the constant "*Jones*" and that variables n_4 and n_5 must represent the same value. The *chase* is an algorithm that transforms tableaux system on the basis of constraints. If the tableaux system in Table 2.17 were chased with respect to $Hospital\_Ward \rightarrow Head$, then the tableaux system in Table 2.19 would be produced.

Table 2.19 Tableaux System

X	Nurse_Name	Hospital_Ward	Head
x_1	Mary	v_d	Jones
x_2	Ann	v_d	Jones
x_3	Betty	n_2	n_3
x_4	v_E	surgical	n_4
x_5	Lucy	surgical	n_5

The variables in tableaux system can be divided into two categories:

- *distinguished variables*, one for each attribute (if b is an attribute of interest, then v_b is the corresponding distinguished variable),
- *nondistinguished variables* (there are countably many of them: $n_1, n_2, \ldots$).

Tableaux system is an information system such that no distinguished variables appear as a value for two distinct attributes.

The notion of containment mapping also requires a preliminary concept. Among the various symbols that may appear as values for an attribute in a tableaux system, we define a partial order, indicated with "$\leq$", as follows:

- constants are pairwise incomparable, and precede all variables (for every constant c and variable v, we have $c \leq v$)
- all distinguished variables precede all non-distinguished variables,
- non-distinguished variables are totally ordered according to the order of their subscripts ($n_i \leq n_j$ iff $i \leq j$).

If $s_1 \leq s_2$ and $s_1 \neq s_2$ then $s_1 < s_2$.

Assume, we have two tableaux systems S_1, S_2 classifying the same sets of objects (let's say objects from X) using the same sets of attributes (e.g. attribute A).

Definition 2.34. *A containment mapping from S_1 to S_2 is a function Ψ from symbols to symbols that satisfies the following conditions:*

- *for every symbol s appearing in S_1, $\Psi(s) \leq s$,*
- *if Ψ is extended to rows in tableaux systems, then for every row r_1 in S_1 there is a row r_2 in S_2 such that $\Psi(r_1(a)) = r_2(a)$).*

Expanding the consequences of the condition $\Psi(s) \leq s$, a containment mapping Ψ maps:

- constants to themselves (the identity on constants),
- each distinguished variable to constant or to itself,
- each non-distinguished variable to a constant, or to the distinguished variable for the corresponding attribute, or to non-distinguished variable with a lower subscript.

We say that tableaux system S_1 represents information system S_2 if there is a containment mapping from S_1 to S_2. For example, the tableaux system in Table 2.17 represents the information system in Table 2.18 because of the containment mapping Ψ given below:

$$\Psi(v_d) = pediatric \qquad \Psi(n_2) = maternity \qquad \Psi(n_1) = Jones$$
$$\Psi(n_3) = Smith \qquad \Psi(n_4) = Brown \qquad \Psi(n_5) = Brown$$

The notion of containment mapping is transitive which means that if there are containment mappings from S_1 to S_2 and from S_2 to S_3, then also there is a containment mapping from S_1 to S_3.

Let us define satisfaction of functional dependencies for tableaux system in the same way as for relations.

Definition 2.35. *A tableaux system S satisfies functional dependency $B \rightarrow C$ if for every pair of rows r_1, r_2 in S such that $r_1[B] = r_2[B]$, it is the case that $r_1[C] = r_2[C]$.*

We can now present the chase algorithm referring to functional dependencies. The algorithm receives as input a tableaux system S and a set of functional dependencies F, with the aim of transforming S into a tableaux system which satisfies F. We say that S is chased with respect to F.

Algorithm CHASE
INPUT

- Tableaux system S,
- Set of functional dependencies F,

BEGIN
$S_1 := S$

while there are $r_1, r_2 \in S_1$ and $(B \to b) \in F$ such that $r_1[B] = r_2[B]$ and $r_1[b] < r_2[b]$ **do**
change all the occurrences of the value $r_2[b]$ in S_1 to $r_1[b]$
$CHASE_F(S) := S_1$
END
OUTPUT

- Tableaux system $CHASE_F(S)$

Algorithm Chase always terminates if applied to a finite tableaux system. If one execution of the algorithm generates a tableaux system that satisfies F, then every execution of the algorithm generates the same tableaux system.

2.9.2 Handling Incomplete Values Using $CHASE_1$ Algorithm

There is a relationship between interpretation of queries and the way the incomplete information in an information system is seen [7], [8], [9].

Assume, for example, that we are concerned with identifying all objects in the system satisfying a given description. For example an information system might contain information about patients in a hospital and classify them using four attributes of *Blood_Pressure*, *Cholesterol_Level*, *Gender* and *Size*. A simple query might be to find all patients with *hypertension* and *low LDL*. When the information system is incomplete, patients having hypertension and unknown cholesterol level can be handled by either including or excluding them from the answer to the query. In the first case we talk about optimistic approach to query interpretation while in the second case we talk about pessimistic approach. Another option to handle such a query would be to discover rules for *Cholesterol_Level* in terms of the attributes *Blood_Pressure*, *Gender*, and *Size*. These rules could then be applied to the patients with unknown cholesterol level to discover this level of LDL and possibly to identify more objects satisfying the query.

Consider that in our example one of the generated rules said:

$$(blood\_pressure, hypertension) * (size, large) \to (LDL, high)$$

Thus, if one of the patients having high blood pressure and large size has no value for cholesterol level, then we should not include this patient in the list of patients with hypertension and low LDL. Attributes *Blood_Pressure* and *Size* are classification attributes while *Cholesterol_Level* is the decision attribute.

Let us give another example showing how close is the relationship between replacing Null values by either incomplete or complete values in an incomplete information system and the way queries are interpreted. Namely, saying that the confidence in object x that he has *Haglund-Sever_Disease* is $\frac{1}{3}$ can be either written as $(H - S\_Disease, \frac{1}{3}) \in Orthopedic\_disease$ or $(x, \frac{1}{3}) \in I(H -$

$S\_Disease$), where I is an interpretation of queries (the term $Brown$ is treated here as a query).

In this section we are interested in how to use rules extracted from an incomplete information system S to replace its null values by values less incomplete. The approach proposed in this book is to use not only functional dependencies to chase system S but also use, for that purpose, rules discovered from a complete subsystem of S. But rules discovered from S do not have to be consistent.

Taking this fact into consideration, the chase algorithm ($Chase_1$) presented in this section allows chasing information system S only with non-conflicting rules. The second chase algorithm called $Chase_2$ has less restrictions and it allows chasing information system S with inconsistent rules as well.

Assume that $S = (X, A, V)$, where $V = \bigcup\{V_a : a \in A\}$ and each $a \in A$ is a partial function from X into $2^{V_a} - \{\emptyset\}$.

In the first step of our algorithm, handling incompleteness in S, all incomplete attributes used in S are identified. An attribute is incomplete if there is an object in S with incomplete information on this attribute. The values of all incomplete attributes in S are treated as concepts to be learned (in a form of rules) either only from S or from S and its remote sites (if S is one of collaborating autonomous information systems).

The second step of the algorithm is to extract rules describing these concepts. These rules are stored in a knowledge base D for information system S (see [35], [36]). The algorithm $Chase_1$ proposed in this section assumes that all inconsistencies in D have to be repaired before they can be used in the chase process. Rules describing attribute value v_a of attribute a are extracted from the subsystem $S_1 = (X_1, A, V)$ of S where $X_1 = \{x \in X : card(a(x)) = 1\}$.

The final step of our procedure is to replace the incomplete information in S by values provided by rules in D. It is done by rule-based chase algorithm $Chase_1$, given below.

To present $Chase_1$ algorithm, we assume first that the set of all incomplete attributes in A is denoted by $In(A)$.
Saying another words $In(A) = \{a \in A : (\exists x \in X)[card(a(x)) \neq 1]\}$.

Definition 2.36. Let $S_j = (X_j, A, V)$ for any j, $1 \leq j \leq k$. Attributes in $S = (X, A, V) = \bigcap\{S_j : 1 \leq j \leq k\}$ are defined as follows :

$$a_S(x) = \bigcap\{a_{S_j}(x) : 1 \leq j \leq k\} \text{ for any } a_S \in A, x \in X.$$

Assume now that $L(D) = \{(t \rightarrow v_c) \in D : c \in In(A)\}$ is a consistent set of rules in S. The algorithm, given below, converts information system S to a new more complete information system $Chase_1(S)$.

Algorithm $CHASE_1(S, In(A), L(D))$
INPUT

- System $S = (X, A, V)$,
- Set of incomplete attributes $In(A) = \{a_1, a_2, \ldots a_k\}$,

BEGIN

$j := 1;$

while $j \leq k$ **do begin**

$S_j := S;$

for all $v \in V_{a_j}$ **do**

while there is $x \in X$ and a rule $(t \rightarrow v) \in L(D)$ such that $x \in N_{S_j}(t)$ and $card(a_j(x)) \neq 1$ **do begin**

$a(x) := v;$ **end**

$j := j + 1;$ **end**

$S := \bigcap \{S_j : 1 \leq j \leq k\}$

$Chase_1(S, In(A), L(D))$

END

OUTPUT

- System $CHASE_1(S)$

The algorithm $Chase_1$ is chasing information system S, attribute by attribute, changing values of attributes assigned to objects in X only after all incomplete attributes in S are being chased. This process is continued till the fixed point is reached (no changes in S are made by $Chase_1$).

Table 2.20 Incomplete Information System

X	Attribute b	Attribute c	Attribute d	Attribute e	Attribute f	Attribute g
x_1	b_1	c_1		e_2	f_1	
x_2	b_2	c_2	d_2	e_1	f_2	g_3
x_3	b_1	c_1	d_3	e_1	f_1	g_1
x_4	b_3	c_3	d_3	e_3	f_1	g_3
x_5	b_2	c_2		e_3	f_1	g_2
x_6		c_1	d_2		f_2	g_1
x_7	b_1		d_2	e_2	f_4	g_1
x_8			d_2	e_2	f_2	g_3
x_9	b_3	c_1	d_1		f_2	
x_{10}	b_2	c_1		e_3	f_4	g_2

Example 2.37. Let us assume that system $S = (X, A, V)$, with

$X = \{x_i\}_{i \in \{1,2,...10\}}$ and $A = \{b, c, d, e, f, g\}$ is represented by Table 2.20.

Clearly, attributes $\{b, c, d, e, g\}$ are incomplete, while the attribute f is complete in system S.

The assumption that $L(D)$ is consistent in $Chase_1$ algorithm is only for simplification purpose but we can easily drop this condition. In this case, before any rule r from $L(D)$ is used by $Chase_1$, it has to be checked if there are no other rules in $L(D)$ which contradict with r. Only then rule r can be used for chasing system S.

To understand the $Chase_1$ better, let us assume that $L(D)$ contains the following rules extracted from S, which define values of attribute b (some rules contradict each other):

$(e, e_2) \to (b, b_1)$ support 2, $(c, c_1) * (f, f_1) \to (b, b_1)$ support 2,
$(g, g_2) \to (b, b_2)$ support 2, $(g, g_3) * (d, d_2) \to (b, b_2)$ support 1,
$(c, c_2) \to (b, b_2)$ support 2, $(e, e_1) * (f, f_1) \to (b, b_1)$ support 1,
$(c, c_3) \to (b, b_3)$ support 1, $(e, e_3) * (d, d_3) \to (b, b_3)$ support 1,
                                      $(f, f_2) * (d, d_2) \to (b, b_2)$ support 1.

There are two null values in S corresponding to attribute b, $b(x_6)$ and $b(x_8)$. Algorithm $Chase_1$ will search for rules which can replace these null values by values of attribute b.

Let us work first on $b(x_6)$. Only one rule:

$$(f, f_2) * (d, d_2) \to (b, b_2) \text{ support 1}$$

can be applied by $Chase_1$. So, in this case $Chase_1$ will change $b(x_6)$ from null value to value b_2. This value is stored in a temporary information system S_1 as $b_{S_1}(x_6) = b_2$.

Now, algorithm $Chase_1$ will try to change the value $b(x_8)$. The following rules can be applied:

$(e, e_2) \to (b, b_1)$ support 2, $(e, e_1) * (f, f_1) \to (b, b_1)$ support 1,
$(g, g_3) * (d, d_2) \to (b, b_2)$ support 1, $(f, f_2) * (d, d_2) \to (b, b_2)$ support 1.

Two rules with a join support 3 propose value b_1 for $b(x_8)$ and two rules with a join support 2 propose value b_2 for $b(x_8)$.

There are two options to construct $L(D)$, either we rule out all contradicting rules and the null value is not being changed or we accept the value which has the highest support. If we follow the second option, we get $b_{S_1}(x_8) = b_1$.

Assume now that $L(D)$ contains the following rules extracted from S which define values of attribute c (some rules contradict each other):

$(b, b_1) \to (c, c_1)$ support 2, $(b, b_2) * (d, d_2) \to (c, c_2)$ support 1,
$(e, e_2) \to (c, c_1)$ support 1, $(b, b_2) * (e, e_1) \to (c, c_2)$ support 1,
$(f, f_4) \to (c, c_1)$ support 1, $(b, b_2) * (f, f_2) \to (c, c_2)$ support 1,
$(g, g_1) \to (c, c_1)$ support 2, $(b, b_2) * (g, g_3) \to (c, c_2)$ support 1,
$(d, d_2) * (e, e_1) \to (c, c_2)$ support 1, $(d, d_2) * (g, g_3) \to (c, c_2)$ support 1.

There are two null values in S corresponding to attribute c, $c(x_7)$ and $c(x_8)$. Algorithm $Chase_1$ will search for rules which can replace these null values by values of attribute c.

Let us work first on $c(x_7)$. The following rules can be applied:

$(b, b_1) \to (c, c_1)$ support 2, $(e, e_2) \to (c, c_1)$ support 1,
$(f, f_4) \to (c, c_1)$ support 1, $(g, g_1) \to (c, c_1)$ support 2.

Since all these rules support the value c_1 for $c(x_7)$, then $c_{S_2}(x_7) = c_1$.

Now, algorithm $Chase_1$ will try to change the value $c(x_8)$. The following rules can be applied:

$(e, e_2) \to (c, c_1)$ support 1, $(d, d_2) * (g, g_3) \to (c, c_2)$ support 1.

Since both rules have the same support, then the value of $c(x_8)$ remains unchanged which means $c_{S_2}(x_8) = V_c$ (undefined).

Assume now that $L(D)$ contains the following rules extracted from S which define values of attribute d (some rules contradict each other):

$(b, b_2) \to (d, d_2)$ support 1, $(e, e_3) \to (d, d_3)$ support 1,
$(c, c_2) \to (d, d_2)$ support 1, $(f, f_1) \to (d, d_3)$ support 2,
$(e, e_2) \to (d, d_2)$ support 2, $(f, f_4) \to (d, d_2)$ support 1,
                                                $(b, b_1) * (c, c_1) \to (d, d_3)$ support 1.

There are three null values in S corresponding to attribute d, $d(x_1)$, $d(x_5)$ and $d(x_{10})$.

Let us work first on $d(x_1)$. The following rules can be applied:

$(e, e_2) \to (d, d_2)$ support 2, $(f, f_1) \to (d, d_3)$ support 2.

Since both rules have the same support, $d_{S_3}(x_1) = V_d$.

Now, let us work on $d(x_5)$. The following rules can be applied:

$(b, b_2) \to (d, d_2)$ support 1, $(e, e_3) \to (d, d_3)$ support 1,
$(c, c_2) \to (d, d_2)$ support 1, $(f, f_1) \to (d, d_3)$ support 2.

Since value d_3 has higher support, $d_{S_3}(x_5) = d_3$.

Finally, we have to work on $d(x_{10})$. The following three rules can be applied:

$(b, b_2) \to (d, d_2)$ support 1, $(e, e_3) \to (d, d_3)$ support 1,
                                  $(f, f_4) \to (d, d_2)$ support 1.

Since d_2 has higher support, then $d_{S_3}(x_{10}) = d_2$.

Assume now that $L(D)$ contains the following rules extracted from S which define values of attribute e (some rules contradict each other):

$(b, b_3) \to (e, e_3)$ support 1, $(c, c_1) * (g, g_1) \to (e, e_1)$ support 1,
                                  $(d, d_2) * (g, g_1) \to (e, e_2)$ support 1.

There are two null values in S corresponding to attribute e: $e(x_6)$, $e(x_9)$.

Let us work first on $e(x_6)$. The following rules can be applied:

$(c, c_1) * (g, g_1) \to (e, e_1)$ support 1, $(d, d_2) * (g, g_1) \to (e, e_2)$ support 1.

It means that $e_{S_4}(x_6) = V_e$.

Now, let us work on $e(x_9)$. Only one rule can be applied:

$$(b, b_3) \rightarrow (e, e_3) \text{ support 1.}$$

It means that $e_{S_4}(x_9) = e_3$.

Assume now that $L(D)$ contains the following rules extracted from S which define values of attribute g (some rules contradict each other):

$(b, b_1) \rightarrow (g, g_1)$ support 2, $(b, b_3) \rightarrow (g, g_3)$ support 1,

$(c, c_1) * (f, f_1) \rightarrow (g, g_1)$ support 1, $(c, c_1) * (f, f_2) \rightarrow (g, g_1)$ support 1.

There are two null values in S corresponding to attribute g, $g(x_1)$, $g(x_9)$.

Let us work first on $g(x_1)$. The following rules can be applied:

$(b, b_1) \rightarrow (g, g_1)$ support 2, $(c, c_1) * (f, f_1) \rightarrow (g, g_1)$ support 1.

It means that $g_{S_5}(x_1) = g_1$.

Now, let us work on $g(x_9)$. The following rules can be applied:

$(b, b_3) \rightarrow (g, g_3)$ support 3, $(c, c_1) * (f, f_2) \rightarrow (g, g_1)$ support 1.

So, $g_{S_5}(x_9) = V_g$.

Table 2.21 Partial Result of Chase on S

X	Attribute b	Attribute c	Attribute d	Attribute e	Attribute f	Attribute g
x_1	b_1	c_1		e_2	f_1	g_1
x_2	b_2	c_2	d_2	e_1	f_2	g_3
x_3	b_1	c_1	d_3	e_1	f_1	g_1
x_4	b_3	c_3	d_3	e_3	f_1	g_3
x_5	b_2	c_2	d_3	e_3	f_1	g_2
x_6	b_2	c_1	d_2		f_2	g_1
x_7	b_1	c_1	d_2	e_2	f_4	g_1
x_8	b_1		d_2	e_2	f_2	g_3
x_9	b_3	c_1	d_1	e_3	f_2	
x_{10}	b_2	c_1	d_2	e_3	f_4	g_2

Now, taking $S = (X, A, V) = \bigcap \{S_j : 1 \leq j \leq k\}$, we get information system S which can be represented by Table 2.21.

The whole process is repeated till no new chased values are identified, which means the procedure $Chase_1$ reaches a fix point.

2.9.3 Handling Incomplete Values Using $CHASE_2$ Algorithm

Using $Chase_1$ algorithm for predicting what attribute value should replace an incomplete value has a clear advantage over any other method for predicting incomplete values, mainly because of the use of existing associations between values of attributes.

To find these associations we can use either any association rule mining algorithm [3] or any rule discovery algorithm like $LERS$ [13] or $Rosetta$ [22]. Unfortunately, these algorithms, including $Chase_1$, do not handle partially incomplete data, where attribute $a(x)$ can have several possible values, for instance, to $\{(a_1, \frac{1}{4}), (a_2, \frac{1}{4}), (a_3, \frac{1}{2})\}$.

Clearly, we assume here that a is an attribute, x is an object, and $\{a_1, a_2, a_3\} \subseteq V_a$. The weights assigned to these three attribute values should be read as:

- the confidence that $a(x) = a_1$ is $\frac{1}{4}$,
- the confidence that $a(x) = a_2$ is $\frac{1}{4}$,
- the confidence that $a(x) = a_3$ is $\frac{1}{2}$.

Now, given two partially incomplete information systems (defined in Definition 2.29) S_1, S_2 classifying the same sets of objects (e.g. from X) using the same sets of attributes (e.g. A), we assume that:

$$a_{S_1}(x) = \{(a_{1i}, p_{1i}) : i \leq m_1\} \text{ and } _{S_2}(x) = \{(a_{2i}, p_{2i}) : i \leq m_2\}.$$

We say that containment relation Ψ holds between S_1 and S_2, if the following two conditions hold:

- $(\forall x \in X)(\forall a \in A)[card(a_{S_1}(x)) \geq card(a_{S_2}(x))]$,
- $(\forall x \in X)(\forall a \in A)[card(a_{S_1}(x)) = card(a_{S_2}(x))] \rightarrow$

$$\rightarrow \sum_{i \neq j} |p_{2i} - p_{2j}| \succ \sum_{i \neq j} |p_{1i} - p_{1j}|.$$

If containment relation Ψ holds between S_1 and S_2 both of type λ, we also say that information system S_1 was mapped onto S_2 by containment mapping Ψ and denote that fact as

$$\Psi(S_1) = S_2$$

which means that

$$(\forall x \in X)(\forall a \in A)[\Psi(a_{S_1}(x)) = \Psi(a_{S_2}(x))].$$

We also say that containment relation Ψ holds between $a_{S_1}(x)$ and $a_{S_2}(x)$, for any $x \in X$, and $a \in A$.

So, the containment mapping Ψ for incomplete information systems does not increase the number of possible attribute values for a given object. If the number of possible attribute values of a given attribute assigned to an object

Table 2.22 Incomplete Information System S_1 of type $\lambda = \frac{1}{4}$ satisfying containment relation

X	Attribute a	Attribute b	Attribute c	Attribute d	Attribute e
x_1	$(a_1,\frac{1}{3}),(a_2,\frac{2}{3})$	$(b_1,\frac{2}{3}),(b_2,\frac{1}{3})$	c_1	d_1	$(e_1,\frac{1}{2}),(e_2,\frac{1}{2})$
x_2	$(a_2,\frac{1}{4}),(a_3,\frac{3}{4})$	$(b_1,\frac{1}{3}),(b_2,\frac{2}{3})$		d_2	e_1
x_3		b_2	$(c_1,\frac{1}{2}),(c_3,\frac{1}{2})$	d_2	e_3
x_4	a_3		c_2	d_1	$(e_1,\frac{2}{3}),(e_2,\frac{1}{3})$
x_5	$(a_1,\frac{2}{3}),(a_2,\frac{1}{3})$	b_1	c_2		e_1
x_6	a_2	b_2	c_3	d_2	$(e_2,\frac{1}{3}),(e_3,\frac{2}{3})$
x_7	a_2	$(b_1,\frac{1}{4}),(b_2,\frac{3}{4})$	$(c_1,\frac{1}{3}),(c_2,\frac{2}{3})$	d_2	e_2
x_8		b_2	c_1	d_1	e_3

Table 2.23 Incomplete Information System S_2 of type $\lambda = \frac{1}{4}$ satisfying containment relation

X	Attribute a	Attribute b	Attribute c	Attribute d	Attribute e
x_1	$(a_1,\frac{1}{3}),(a_2,\frac{2}{3})$	$(b_1,\frac{2}{3}),(b_2,\frac{1}{3})$	c_1	d_1	$(e_1,\frac{1}{2}),(e_2,\frac{1}{2})$
x_2	$(a_2,\frac{1}{4}),(a_3,\frac{3}{4})$	b_2	$(c_1,\frac{1}{3}),(c_2,\frac{2}{3})$	d_2	e_1
x_3	a_1	b_2	$(c_1,\frac{1}{2}),(c_3,\frac{1}{2})$	d_2	e_3
x_4	a_3		c_2	d_1	e_2
x_5	$(a_1,\frac{2}{3}),(a_2,\frac{1}{3})$	b_1	c_2		e_1
x_6	a_2	b_2	c_3	d_2	$(e_2,\frac{1}{3}),(e_3,\frac{2}{3})$
x_7	a_2	$(b_1,\frac{1}{4}),(b_2,\frac{3}{4})$	c_1	d_2	e_2
x_8	$(a_1,\frac{2}{3}),(a_2,\frac{1}{3})$	b_2	c_1	d_1	e_3

is not changed, then the average difference in confidence assigned to these attribute values has to increase.

Let us take into consideration two systems S_1, S_2 of type $\lambda = \frac{1}{4}$ (Table 2.22 and Table 2.23). It can be easily checked that the values assigned to $a(x_3), a(x_8), b(x_2), c(x_2), c(x_7), e(x_4)$ in S_1 have been changed in S_2.

In each of these six cases, a new attribute value assigned to an object in S_2 is less general than in S_1. It means that $\Psi(S_1) = S_2$.

Assume now that $L(D) = \{(t \to v_c) \in D : c \in In(A)\}$ is the set of all rules extracted from S by $ERID(S, \lambda_1, \lambda_2)$, where λ_1, λ_2 are thresholds respectively for minimum support and minimum confidence.

$ERID$ is the algorithm for discovering rules from incomplete information systems which was presented in Section 2.8.5.

The new algorithm, given below, converts information system S of type λ to a new, more complete information system $CHASE_2(S)$.

Algorithm $CHASE_2(S, In(A), L(D))$
INPUT

- System $S = (X, A, V)$,
- Set of incomplete attributes $In(A) = \{a_1, a_2, \ldots a_k\}$,
- Set of rules $L(D)$

BEGIN
$j := 1$;
while $j \leq k$ **do begin**
$S_j := S$;
for all $x \in X$ **do**
$p_j := 0$;
begin
$b_j(x) := \emptyset$, $n_j := 0$;
for all $v \in V_{a_j}$ **do begin**
if $card(a_j(x)) \neq 1$ and $\{(t_i \rightarrow v) : i \in I\}$ is a maximal subset of rules from
$L(D)$ such that $(x, p_i) \in N_{S_j}(t_i)$ **then**
if $\sum_{i \in I}[p_i \cdot conf(t_i \rightarrow v) \cdot sup(t_i \rightarrow v)] \geq \lambda$ **then begin**
$b_j(x) := b_j(x) \cup \{(v, \sum_{i \in I}[p_i \cdot conf(t_i \rightarrow v) \cdot sup(t_i \rightarrow v)])\}$;
$n_j := n_j + \sum_{i \in I}[p_i \cdot conf(t_i \rightarrow v) \cdot sup(t_i \rightarrow v)]$; **end; end**
$p_j := p_j + n_j$; **end**
if $\Psi(a_j(x)) = [b_j(x)/p_j]$; (containment relation holds between $a_j(x), b_j(x)/p_j$)
then $a_j(x) := [b_j(x)/p_j]$
$j := j + 1$; **end**
$S := \bigcap\{S_j : 1 \leq j \leq k\}$
$CHASE_2(S, In(A), L(D))$
END
OUTPUT

- System $CHASE_2(S)$

Definition of system S is similar to the one in Section 2.9.2. To define system S it is enough to assume that:

$$a_S(x) = (if\, a = a_j \text{ then } a_{S_j}(x)) \text{ for any attribute } a \text{ and object } x.$$

Also, if

$$b_j(x) = \{(v_i, p_i)\}_{i \in I} \text{ then } [b_j(x)/p] = \{(v_i, p_i/p)\}_{i \in I}.$$

Let us apply $CHASE_2$ instead of $CHASE_1$ to the information system given in Table 2.23. We only show how values of the attribute e will change. Similar process is applied to all incomplete attributes in S. After changes of all incomplete attributes are recorded, system S is replaced by $\Psi(S)$ and the whole process is recursively repeated till the fix point is reached.

Example 2.38. Assume that partially incomplete information system S is represented by Table 2.24. Here we have $X = \{x_1, x_2, x_3, x_4, x_5, x_6, x_7, x_8\}$ and $A = \{a, b, c, d, e\}$.

Table 2.24 Incomplete Information System S_1 of type $\lambda = 0.3$

X	Attribute a	Attribute b	Attribute c	Attribute d	Attribute e
x_1	$(a_1, \frac{1}{3}), (a_2, \frac{2}{3})$	$(b_1, \frac{2}{3}), (b_2, \frac{1}{3})$	c_1	d_1	$(e_1, \frac{1}{2}), (e_2, \frac{1}{2})$
x_2	$(a_2, \frac{1}{4}), (a_3, \frac{3}{4})$	$(b_1, \frac{1}{3}), (b_2, \frac{2}{3})$		d_2	e_1
x_3	a_1	b_2	$(c_1, \frac{1}{2}), (c_3, \frac{1}{2})$	d_2	e_3
x_4	a_3		c_2	d_1	$(e_1, \frac{2}{3}), (e_2, \frac{1}{3})$
x_5	$(a_1, \frac{2}{3}), (a_2, \frac{1}{3})$	b_1	c_2		e_1
x_6	a_2	b_2	c_3	d_2	$(e_2, \frac{1}{3}), (e_3, \frac{2}{3})$
x_7	a_2	$(b_1, \frac{1}{4}), (b_2, \frac{3}{4})$	$(c_1, \frac{1}{3}), (c_2, \frac{2}{3})$	d_2	e_2
x_8	a_3	b_2	c_1	d_1	e_3

Let us try to extract rules from S describing attribute e in terms of attributes $\{a, b, c, d\}$ following a strategy similar to *LERS*.

First we start with identifying sets of objects in X having the properties $\{a_1, a_2, a_3, b_1, b_2, c_1, c_2, c_3, d_1, d_2\}$ and next we will find their relationships with sets of objects in X having properties $\{e_1, e_2\}$ and $\{e_3\}$.

For simplicity in this example we will use notation a_1 instead of (a, a_1). Let us start with a_1^* which will be equal to $\{(x_1, \frac{1}{3}), (x_3, 1), (x_5, \frac{2}{3})\}$. The justification of this fact is quite simple. Only these three objects may have that property. Object x_3 has property a_1 with confidence 1 for sure. The confidence that x_1 has property a_1 is $\frac{1}{3}$, since $(a_1, \frac{1}{3}) \in a(x_1)$. In a similar way we justify property a_1 for object x_5 (with confidence $\frac{2}{3}$) So, as far as values of classification attributes, we get:

$$a_1^* = \{(x_1, \frac{1}{3}), (x_3, 1), (x_5, \frac{2}{3})\}$$
$$a_2^* = \{(x_1, \frac{2}{3}), (x_2, \frac{1}{4}), (x_5, \frac{1}{3}), (x_6, 1), (x_7, 1)\}$$
$$a_3^* = \{(x_2, \frac{3}{4}), (x_4, 1), (x_8, 1)\}$$
$$b_1^* = \{(x_1, \frac{2}{3}), (x_2, \frac{1}{3}), (x_4, \frac{1}{2}), (x_5, 1), (x_7, \frac{1}{4})\}$$
$$b_2^* = \{(x_1, \frac{1}{3}), (x_2, \frac{2}{3}), (x_3, 1), (x_4, \frac{1}{2}), (x_6, 1), (x_7, \frac{3}{4}), (x_8, 1)\}$$
$$c_1^* = \{(x_1, 1), (x_2, \frac{1}{3}), (x_3, \frac{1}{2}), (x_7, \frac{1}{3}), (x_8, 1)\}$$
$$c_2^* = \{(x_2, \frac{1}{3}), (x_4, 1), (x_5, 1, (x_7, \frac{2}{3})\}$$
$$c_3^* = \{(x_2, \frac{1}{3}), (x_3, \frac{1}{2}), (x_6, 1)\}$$
$$d_1^* = \{(x_1, 1), (x_4, 1), (x_5, \frac{1}{2}), (x_8, 1)\}$$
$$d_2^* = \{(x_2, 1), (x_3, 1), (x_5, \frac{1}{2}), (x_6, 1), (x_7, 1)\}.$$

For the values of decision attribute we get:

$$e_1^* = \{(x_1, \frac{1}{2}), (x_2, 1), (x_4, \frac{2}{3}), (x_5, 1)\}$$
$$e_2^* = \{(x_1, \frac{1}{2}), (x_4, \frac{1}{3}), (x_6, \frac{1}{3}), (x_7, 1)\}$$
$$e_3^* = \{(x_3, 1), (x_6, \frac{2}{3}), (x_8, 1)\}.$$

The next step is to propose a method for checking a relationship between classification attributes and the decision attribute.

For any two sets $c_i^* = \{x_i, p_i\}_{i \in N}$, $e_j^* = \{y_j, q_j\}_{j \in M}$, where $p_i > 0$ and $q_j > 0$, we propose that: $\{x_i, p_i\}_{i \in N} \subseteq \{y_j, q_j\}_{j \in M}$ iff the support of the rule

$c_i \to e_j$ is above some threshold value. So, the relationship between c_i^*, e_j^* depends only on how high is the support of the corresponding rule $c_i \to e_j$.

Coming back to our example, let us notice that we may have a successful set-theoretical inclusion between objects from a_1^* and objects from e_3^* but the set-theoretical inclusion between objects from a_1^* and objects from e_3^* may fail. The justification of this claim is the following.

Object x_3 is a member of a_1^* but it does not belong to e_1^*. At the same time, the other set-theoretical inclusion is feasible because it is not certain that x_1 and x_5 belong to a_1^*. Assuming that the relationship $a_1^* \subseteq e_3^*$ is successful, what can we say about the support and confidence of the rule $a_1 \to e_3$? Object x_1: supports a_1 with a confidence $\frac{1}{3}$, and e_3 with a confidence 0. Object x_3: supports a_1 with a confidence 1, and e_3 with a confidence 1. Object x_5: supports a_1 with a confidence $\frac{2}{3}$, and e_3 with a confidence 0.

We calculate the support of the rule $a_1 \to e_3$ in following way:

$$\frac{1}{3} \cdot 0 + 1 \cdot 1 + \frac{2}{3} \cdot 0 = 1.$$

Similarly, the support of the term a_1 is equal $\frac{1}{3} + 1 + \frac{2}{3} = 2$. Therefore, the confidence of the above rule will be $\frac{1}{2}$, if we follow the standard strategy for calculating the confidence of a rule.

Now, let us follow a strategy which has some similarity with *LERS*. All inclusions " $\subseteq$ " will be automatically marked assuming that the confidence and support of the corresponding rules are both above some threshold values (which are given by user). In the current example we take 1 as the threshold for support and $\frac{1}{2}$ as the threshold of confidence.

Assume again that $c_i^* = \{x_i, p_i\}_{i \in N}$, $e_j^* = \{y_j, q_j\}_{j \in M}$. The algorithm will check first the support of the rule $c_i \to e_j$. If support is below a threshold value, then the corresponding relationship $\{x_i, p_i\}_{i \in N} \subseteq \{y_j, q_j\}_{j \in M}$ does not hold and it will be not considered in later steps. Otherwise, the confidence of the rule $c_i \to e_j$ is checked. If that confidence is either above or equals the assumed threshold value, the rule is approved and the corresponding relationship $\{x_i, p_i\}_{i \in N} \subseteq \{y_j, q_j\}_{j \in M}$ is marked. Otherwise this corresponding relationship remains unmarked.

$a_1^* \subseteq e_1^*$ $(sup = \frac{5}{6} < 1$)   - *marked negative*
$a_1^* \subseteq e_2^*$ $(sup = \frac{1}{6} < 1$)   - *marked negative*
$a_1^* \subseteq e_3^*$ $(sup = 1 \geq 1$ and $conf = 0.5)$ - **marked positive**

$a_2^* \subseteq e_1^*$ $(sup = \frac{11}{12} < 1$)   - *marked negative*
$a_2^* \subseteq e_2^*$ $(sup = \frac{5}{3} \geq 1$ and $conf = 0.51$) - **marked positive**
$a_2^* \subseteq e_3^*$ $(sup = \frac{2}{3} < 1)$ - *marked negative*

$a_3^* \subseteq e_1^*$ $(sup = \frac{17}{12} \geq 1$ and $conf = 0.51)$ - **marked positive**
$a_3^* \subseteq e_2^*$ $(sup = \frac{1}{3} < 1)$ - *marked negative*
$a_3^* \subseteq e_3^*$ $(sup = 1 \geq 1$ but $conf = 0.36)$ - not marked

$b_1^* \subseteq e_1^*$ $(sup = 2 \geq 1$ and $conf = 0.72)$ — ***marked positive***
$b_1^* \subseteq e_2^*$ $(sup = \frac{3}{4} < 1$) — *marked negative*
$b_1^* \subseteq e_3^*$ $(sup = 0 < 1)$ — *marked negative*

$b_2^* \subseteq e_1^*$ $(sup = \frac{7}{6} \geq 1$ but $conf = 0.22)$ — not marked
$b_2^* \subseteq e_2^*$ $(sup = \frac{17}{12} \geq 1$ but $conf = 0.27$) — not marked
$b_2^* \subseteq e_3^*$ $(sup = \frac{8}{3} \geq 1$ and $conf = 0.51)$ — ***marked positive***

$c_1^* \subseteq e_1^*$ $(sup = \frac{5}{6} < 1$) — *marked negative*
$c_1^* \subseteq e_2^*$ $(sup = \frac{5}{6} < 1$) — *marked negative*
$c_1^* \subseteq e_3^*$ $(sup = \frac{3}{2} \geq 1$ but $conf = 0.47)$ — not marked

$c_2^* \subseteq e_1^*$ $(sup = 2 \geq 1$ and $conf = 0.66)$ — ***marked positive***
$c_2^* \subseteq e_2^*$ $(sup = 1 \geq 1$ but $conf = 0.33$) — - not marked
$c_2^* \subseteq e_3^*$ $(sup = 0 < 1)$ — *marked negative*

$c_3^* \subseteq e_1^*$ $(sup = \frac{1}{3} < 1)$ — *marked negative*
$c_3^* \subseteq e_2^*$ $(sup = \frac{1}{3} < 1)$ — *marked negative*
$c_3^* \subseteq e_3^*$ $(sup = \frac{7}{6} \geq 1$ and $conf = 0.64)$ — ***marked positive***

$d_1^* \subseteq e_1^*$ $(sup = \frac{5}{3} \geq 1$ but $conf = 0.48)$ — not marked
$d_1^* \subseteq e_2^*$ $(sup = \frac{5}{6} < 1$) — *marked negative*
$d_1^* \subseteq e_3^*$ $(sup = 1 \geq 1$ but $conf = 0.28)$ — not marked

$d_2^* \subseteq e_1^*$ $(sup = \frac{3}{2} \geq 1$ but $conf = 0.33)$ — not marked
$d_2^* \subseteq e_2^*$ $(sup = \frac{1}{3} < 1)$ — *marked negative*
$d_2^* \subseteq e_3^*$ $(sup = \frac{5}{3} \geq 1$ but $conf = 0.37)$ — not marked

We obtained seven marked positive rules, where support and threshold hold, and nine rules, where only support holds. The next step is to build terms of pairs from objects with support greater or equal 1, and confidence less than 0.5.

We propose the following definition for concatenating any two sets $c_i^* = \{x_i, p_i\}_{i \in N}$, $e_j^* = \{y_j, q_j\}_{j \in M}$ where $K = M \cap N : (c_i \cdot e_j)^* = \{(x_i, p_i \cdot q_j)\}_{i \in K}$.

Following this definition, we get:

$(a_3, c_1)^* \subseteq e_3^*$ $(sup = 1 \geq 1$ and $conf = 0.8$) — ***marked positive***
$(a_3, d_1)^* \subseteq e_3^*$ $(sup = 1 \geq 1$ and $conf = 0.5$) — ***marked positive***
$(a_3, d_2)^* \subseteq e_3^*$ $(sup = 0 < 1$) — *marked negative*

$(b_2, d_1)^* \subseteq e_1^* \ (sup = \frac{1}{6} < 1)$ *- marked negative*
$(b_2, d_2)^* \subseteq e_1^* \ (sup = \frac{2}{3} < 1)$ *- marked negative*
$(b_2, c_2)^* \subseteq e_2^* \ (sup = \frac{2}{3} < 1)$ *- marked negative*

$(c_1, d_1)^* \subseteq e_3^* \ (sup = 1 \geq 1$ and $conf = 0.5$)   **- marked positive**
$(c_1, d_2)^* \subseteq e_3^* \ (sup = \frac{1}{2} < 1)$ *- marked negative*

All sets were marked, therefore we can already start to extract rules from an incomplete information system. Applying algorithm $ERID(S, \lambda_1, \lambda_2)$ to system S, from Table 2.24, with two thresholds: $\lambda_1 = 1, \lambda_2 = 0.5$, we get ten following rules:

$$r_1 = a_1 \rightarrow e_3, \qquad sup(r_1) = 1, \quad conf(r_1) = 0.5$$
$$r_2 = a_2 \rightarrow e_2, \qquad sup(r_2) = \frac{5}{3}, \quad conf(r_1) = 0.51$$
$$r_3 = a_3 \rightarrow e_1, \qquad sup(r_3) = \frac{17}{12}, \quad conf(r_3) = 0.51$$
$$r_4 = b_1 \rightarrow e_1, \qquad sup(r_4) = 2, \quad conf(r_4) = 0.72$$
$$r_5 = b_2 \rightarrow e_3, \qquad sup(r_5) = \frac{8}{3}, \quad conf(r_5) = 0.51$$
$$r_6 = c_2 \rightarrow e_1, \qquad sup(r_6) = 2, \quad conf(r_6) = 0.66$$
$$r_7 = c_3 \rightarrow e_3, \qquad sup(r_7) = \frac{7}{6}, \quad conf(r_7) = 0.64$$
$$r_8 = a_3 * c_1 \rightarrow e_3, \qquad sup(r_8) = 1, \quad conf(r_8) = 0.8$$
$$r_9 = a_3 * d_1 \rightarrow e_3, \qquad sup(r_9) = 1, \quad conf(r_9) = 0.5$$
$$r_{10} = c_1 * d_1 \rightarrow e_3, \qquad sup(r_{10}) = 1, \quad conf(r_{10}) = 0.5$$

Only two values $e(x_1)$, $e(x_6)$ of the attribute e can be changed. Below we show how to compute these two values and decide if the current attribute values assigned to objects x_1, x_6 can be replaced by them.

Similar process is applied to all incomplete attributes in S. After all changes of all incomplete attributes are recorded, system S is replaced by and the whole process is recursively repeated till some fix point is reached.

Algorithm $CHASE_2$ will try to replace the current value of $e(x_1) = \{(e_1, \frac{1}{2}), (e_2, \frac{1}{2})\}$ by a new value $e_{new}(x_1)$, initially denoted by the term $\{(e_1, ?), (e_2, ?), (e_3, ?)\}$.

We will show that $\Psi(e(x_1)) = e_{new}(x_1)$, which means that the value $e(x_1)$ will be changed by $CHASE_2$.

To justify our claim, let us compute $e_{new}(x)$ for $x \in \{x_1, x_4, x_6\}$.

For x_1:

$$(e_3, \frac{1}{2} \cdot 1 \cdot \frac{1}{2} + \frac{1}{3} \cdot \frac{8}{3} \cdot \frac{51}{100} + 1 \cdot 1 \cdot \frac{1}{2}) = (e_3, 1.119)$$
$$(e_2, \frac{2}{3} \cdot \frac{5}{3} \cdot \frac{51}{100}) = (e_2, 1.621)$$
$$(e_1, \frac{2}{3} \cdot 2 \cdot \frac{72}{100}) = (e_1, 0.96)$$

we have:

$$e_{new}(x_1) = \{(e_1, \frac{0.96}{0.96 + 1.621 + 1.119}),$$

$$(e_2, \frac{1.621}{0.96 + 1.621 + 1.119}), (e_3, \frac{1.119}{0.96 + 1.621 + 1.119})\} =$$

$$= \{(e_1, 0.26), (e_2, 0.44), (e_3, 0.30)\}.$$

Because, the confidence assigned to e_1 is below the threshold λ, then only two values remain:

$$e_{new}(x_1) = \{(e_2, 0.44), (e_3, 0.30)\}.$$

So, the new value of attribute e assigned to x_1 is

$$e_{new}(x_1) = \{(e_2, \frac{0.44}{0.44+0.30}), (e_3, \frac{0.30}{0.44+0.30})\} = \{(e_2, 0.58), (e_3, 0.42)\}$$

For x_4:

$(e_3, 1 \cdot 1 \cdot \frac{1}{2}) = (e_3, 0.5)$
$(e_2, 0)$
$(e_1, 1 \cdot \frac{17}{12} \cdot \frac{51}{100}) = (e_1, 0.7225)$

we have:

$$e_{new}(x_4) = \{(e_1, \frac{0.7225}{0.5 + 0.7225}), (e_3, \frac{0.5}{0.5 + 0.7225})\} = \{(e_1, 0.59), (e_3, 0.41)\}.$$

So, the new value of attribute e assigned to x_4 remains unchanged. For x_6:

$(e_3, \frac{8}{3} \cdot 1 \cdot \frac{51}{100} + 1 \cdot \frac{7}{6} \cdot \frac{64}{100}) = (e_3, 2.11)$
$(e_2, 1 \cdot \frac{5}{3} \cdot \frac{51}{100}) = (e_2, 0.85)$
$(e_1, 0)$

we have:

$$e_{new}(x_6) = \{(e_2, \frac{0.85}{2.11 + 0.85}), (e_3, \frac{2.11}{2.11 + 0.85})\} = \{(e_2, 0.29), (e_3, 0.71)\}.$$

Because, the confidence assigned to e_3 is below the threshold λ, then only one value remains:

$$e_{new}(x_6) = \{(e_3, 0.71)\}.$$

So, the new value of attribute e assigned to x_6 is e_3.

Table 2.25 Resulting Information System

X	Attribute a	Attribute b	Attribute c	Attribute d	Attribute e
x_1	$(a_1, \frac{1}{3}), (a_2, \frac{2}{3})$	$(b_1, \frac{2}{3}), (b_2, \frac{1}{3})$	c_1	d_1	$(e_2, 0.59), (e_3, 0.41)$
x_2	$(a_2, \frac{1}{4}), (a_3, \frac{3}{4})$	$(b_1, \frac{1}{3}), (b_2, \frac{2}{3})$		d_2	e_1
x_3	a_1	b_2	$(c_1, \frac{1}{2}), (c_3, \frac{1}{2})$	d_2	e_3
x_4	a_3		c_2	d_1	$(e_1, \frac{2}{3}), (e_2, \frac{1}{3})$
x_5	$(a_1, \frac{2}{3}), (a_2, \frac{1}{3})$	b_1	c_2		e_1
x_6	a_2	b_2	c_3	d_2	e_3
x_7	a_2	$(b_1, \frac{1}{4}), (b_2, \frac{3}{4})$	$(c_1, \frac{1}{3}), (c_2, \frac{2}{3})$	d_2	e_2
x_8	a_3	b_2	c_1	d_1	e_3

The new resulting information system can have form as in Table 2.25.

The example presented above shows that the values of attributes stored in the resulting table depend on the threshold λ. Smaller the threshold λ, the level of incompleteness of the information system will get lower.

3

Action Rules

There are two frameworks for generating action rules: loosely coupled and tighty coupled [21]. The loosely coupled framework is often called rule-based. It is based on pairing certain classification rules which have to be discovered first by using for instance algorithms such as LERS [13] or ERID [5], [7]. The tightly coupled framework is often called object-based and it assumes that action rules are discovered directly from a database [6], [16], [18]. Classical methods for discovering them follow algorithms either based on frequent sets (called action sets) and association rules mining [3] or they use algorithms such as LERS or ERID with atomic action sets used as their starting step.

3.1 Main Assumptions

Action rules, introduced by Ras and Wieczorkowska [52] may be utilized by any type of industry maintaining large databases, especially medical, military, education and business. They are constructed from classification rules which suggest ways to re-classify objects, such as patients, students or customers to a desired state. But, very often, such a change cannot be done directly to a chosen attribute (for instance to the attribute *pressure* or *profit*). There-fore, in a natural way, there comes a need to learn definitions of such an attribute in terms of other attributes. Taking into consideration a medical database, these definitions are used then to construct action rules showing what changes in values of attributes, for a given patient are needed, in order to re-classify this patient from one group to another, suggested by doctors. This re-classification may mean that a doctor not interested in a certain method of his patient's treatment, now may use it, and may shift the patient from hospitalized persons into a group of patients who can stay at home. These groups are described by values of classification attributes in a decision system schema.

Let us start with definitions of action terms, action rules and their standard interpretation.

A. Dardzinska: *Action Rules Mining*, SCI 468, pp. 47–89.
DOI: 10.1007/978-3-642-35650-6_3   © Springer-Verlag Berlin Heidelberg 2013

Definition 3.1. *By an atomic action term we mean an expression* $(a, a_1 \to a_2)$, *where* a *is an attribute, and* $\{a_1, a_2\} \in V_a$.

If $a_1 = a_2$ then a is called stable on a_1. In this case, for simplicity reason, we use notation (a, a_1) instead of $(a, a_1 \to a_2)$.

Definition 3.2. *By a set of action terms we mean the smallest set such that:*

1. *If* t *is an atomic action term, then* t *is an action term.*
2. *If* t_1, t_2 *are action terms, then* $t_1 * t_2$ *is an action term.*
3. *If* t *is an action term containing* $(a, a_1 \to a_2)$, $(b, b_1 \to b_2)$ *as its subterms, then* $a \neq b$.

Definition 3.3. *By the domain of an action term* t, *denoted by* $Dom(t)$, *we mean the set of all attribute names listed in* t.

Definition 3.4. *By an action rule we mean an expression* $r = [t_1 \to t_2]$, *where* t_1 *is an action term, and* t_2 *is an atomic action term.*

Additionally we assume, that $Dom(t_2) = \{d\}$ and $Dom(t_1) \subseteq A$, where A is a set of attributes.
The domain $Dom(r)$ of action rule r is defined as $Dom(t_1) \cup Dom(t_2)$.
   Now, let us give an example of action rule.

Table 3.1 Information system S

X	Attribute a	Attribute b	Attribute c	Attribute d
x_1	a_1	b_1	c_1	H
x_2	a_2	b_2	c_1	H
x_3	a_2	b_1	c_1	A
x_4	a_1	b_1	c_2	A
x_5	a_2	b_1	c_2	A
x_6	a_2	b_2	c_2	H
x_7	a_1	b_1	c_2	A
x_8	a_1	b_2	c_1	A
x_9	a_1	b_1	c_1	H
x_{10}	a_2	b_2	c_1	H

Example 3.5. Assume that a decision system is represented as Table 3.1. Expressions

$$(a, a_2 \to a_2), (b, b_2 \to b_1), (c, c_2 \to c_2), (c, c_3 \to c_3), (d, H \to A)$$

are examples of atomic action terms.
Expressions

$$(a, a_2 \to a_2) = (a, a_2), (c, c_2 \to c_2) = (c, c_2), (c, c_3 \to c_3) = (c, c_3)$$

mean that the values respectively a_2, c_2, c_3 of attributes a and c remain unchanged, while $(b, b_2 \to b_1)$ means that the value of attribute b is changed from b_2 to b_1.
Expressions

$$r_1 = [((a, a_2) * (b, b_2 \to b_1)) \to (d, H \to A)],$$

$$r_2 = [[(c, c_2) * (b, b_2 \to b_1)] \to (d, H \to A)]$$

are the examples of action rules. The rule r_1 says that if the value a_2 remains unchanged and value b will change from b_2 to b_1 for a given object x, then it is expected that the value d will change from H to A for object x. Clearly, $Dom(r_1) = \{a, b, d\}$. In a similar way, the rule r_2 says that if the value c_2 remains unchanged and value b will change from b_2 to b_1, then it is expected that the value d will change from H to A, and $Dom(r_2) = \{b, c, d\}$.

Definition 3.6. *Standard interpretation N_S of action terms in $S = (X, A, V)$ is defined as follow:*

1. *If $(a, a_1 \to a_2)$ is an atomic term, then $N_S((a, a_1 \to a_2)) = [\{x \in X : a(x) = a_1\}, \{x \in X : a(x) = a_2\}]$*
2. *If $t_1 = (a, a_1 \to a_2) * t$ and $N_S(t) = [Y_1, Y_2]$, then $N_S(t_1) = [Y_1 \cap \{x \in X : a(x) = a_1\}, Y_2 \cap \{x \in X : a(x) = a_2\}]$.*

Now let us define $[Y_1, Y_2] \cap [Z_1, Z_2]$ as $[Y_1 \cap Z_1, Y_2 \cap Z_2]$ and assume that $N_S(t_1) = [Y_1, Y_2]$ and $N_S(t_2) = [Z_1, Z_2]$. Then $N_S(t_1 * t_2) = N_S(t_1) \cap N_S(t_2)$. Let $r = [t_1 \to t_2]$ be an action rule, where $N_S(t_1) = [Y_1, Y_2]$, $N_S(t_2) = [Z_1, Z_2]$.

Definition 3.7. *By the support and confidence of rule r we mean:*

1. *$sup(r) = min\{card(Y_1 \cap Z_1), card(Y_2 \cap Z_2)\}$*
2. *$conf(r) = \frac{card(Y_1 \cap Z_1)}{card(Y_1)} \cdot \frac{card(Y_2 \cap Z_2)}{card(Y_2)}$ if $card(Y_1) \neq 0, card(Y_2) \neq 0$, $card(Y_1 \cap Z_1) \neq 0$, $card(Y_2 \cap Z_2) \neq 0$*
3. *$conf(r) = 0$ in other cases*

The definition of a confidence should be interpreted as an optimistic confidence. Coming back to the Example 3.5, we can find many action rules associated with system S. Let us take two rules extracted earlier:

$r_1 = [((a, a_2) * (b, b_2 \to b_1)) \to (d, H \to A)],$
$r_2 = [((c, c_2) * (b, b_2 \to b_1)) \to (d, H \to A)].$

For rule r_1 we have:

$N_S((a, a_2 \to a_2)) = [\{x_2, x_3, x_5, x_6, x_{10}\}, \{x_2, x_3, x_5, x_6, x_{10}\}]$
$N_S((b, b_2 \to b_1)) = [\{x_2, x_6, x_8, x_{10}\}, \{x_1, x_3, x_4, x_5, x_7, x_9\}]$
$N_S((a, a_2 \to a_2) * (b, b_2 \to b_1)) = [\{x_2, x_6, x_{10}\}, \{x_3, x_5\}]$
$N_S((d, H \to A)) = [\{x_1, x_2, x_6, x_9, x_{10}\}, \{x_3, x_4, x_5, x_7, x_8\}]$

Therefore the support and confidence are $sup(r_1) = 2$, $conf(r_1) = \frac{3}{3} \cdot \frac{2}{2} = 1$. For rule r_2 we have:

$$N_S((c, c_2 \to c_2)) = [\{x_4, x_5, x_6, x_7\}, \{x_4, x_5, x_6, x_7\}]$$
$$N_S((b, b_2 \to b_1)) = [\{x_2, x_6, x_8, x_{10}\}, \{x_1, x_3, x_4, x_5, x_7, x_9\}]$$
$$N_S((c, c_2 \to c_2) * (b, b_2 \to b_1)) = [\{x_6\}, \{x_4, x_5, x_7\}]$$
$$N_S((d, H \to A)) = [\{x_1, x_2, x_6, x_9, x_{10}\}, \{x_3, x_4, x_5, x_7, x_8\}]$$

Therefore the support and confidence are $sup(r_1) = 2$, $conf(r_1) = 1 \cdot \frac{3}{3} = 1$.

3.2 Action Rules from Classification Rules

Finding useful rules is a very important and extremely interesting task of knowledge discovery in data. Most of the researchers focused on techniques for generating patterns, such as classification rules or association rules, from data sets. They assume that the user should analyze these patterns and infer actionable solutions for specific problems within given domains. The classical knowledge discovery algorithms have the potential to identify enormous number of significant patterns from data. Therefore, people are overwhelmed by a large number of uninteresting patterns which are very difficult to analyze and give time consuming solutions. So, there is still a need to look for new methods and tools with the ability to assist people in identifying rules with useful knowledge. There are two types of interestingness measure: subjective and objective [1], [26], [57]. An objective measure is a data-driven approach for evaluating the quality of association patterns. It is domain-independent and requires minimal input from the users, other than to specify a threshold for filtering low-quality patterns [59]. An objective measure is usually computed based on the frequency counts tabulated in a contingency table.Subjective interestingness measures include actionability [1] and unexpectedness [57]. When a rule contradicts the user's prior belief about the domain, surprises him or uncovers new knowledge, it is classified as unexpected. A rule is deemed actionable, if the user can take action to gain an advantage based on this rule. Domain experts basically look at a rule and say that this rule can be converted into an appropriate action. E-action rules mining is a method which helps people in intelligently and automatically way to acquire useful information from data. This information can be turned into actions. The approach gives suggestions about how to change certain attribute values of a given set of objects in order to reclassify them according to a user wish. Two frameworks for mining actionable knowledge can be taken into consideration: rule-based and object-based [21]. In the object-based approach, action rules are extracted directly from a database [6], [16], [18] ,while in rule-based approach [52], the extraction of actionable knowledge is a consequence of using classification rules discovery. It is further subdivided into: methods generating action rules from certain pairs of classification rules [50], [51], [63], [66], and methods generating action rules from single classification rules [40], [48]. For example algorithm ARAS, proposed in [51] generates sets of terms (built from values of attributes) around classification rules and constructs action rules directly from them. In most of the algorithms for action rules mining, there is no guarantee that the discovered patterns in the first step will lead to actionable knowledge that is capable of maximizing profits. One way to approach this problem is to assign a cost function to all changes of attribute values [64]. If changes of attribute values in the classification part of an action rule are too complex, then they can be replaced by composing such rule with other action rules, as proposed in [63]. Each composition of these rules uniquely defines a new action rule. Objects supporting each new action rule are the same as objects supporting the action rule replaced by it, but the cost of reclassifying them is lower for the new rule. E-action rule forms the actionability concept in a better way than action rule [52] by introducing a notion of its supporting class of objects. E-action rules are constructed from certain pairs of classification rules. They can be used not only for evaluating discovered patterns but also for reclassifying some objects in a dataset from one state into a new more desired one.

For example, classification rules found from a bakery's data can be very useful to describe who is a good client (whom to offer some additional promotions) and who is a bad client (whom to watch to minimize loses). However, if shop managers need to improve their understanding of customers and seek for specific actions to improve the services, mere classification rules are not sufficient. We suggest using classification rules for introducing a new method connected with action based on their condition features in order to get a desired effect on their decision feature. When we look the bakery example again, the strategy of action would consist of modifying some condition features in order to improve our understanding of customers behavior and then improve the services.

E-action rules are useful in many other fields, including medical diagnosis. In medical diagnosis, e.g. in children flat foot problem, classification rules can explain the relationships between symptoms and sickness and help to predict the diagnosis of a new patient. E-action rules are useful in providing a hint to a doctor what symptoms have to be modified or eliminated in order to recover a certain group of patients with better prognoses in their illness.

3.2.1   System DEAR

Proposed algorithm, implemented as system DEAR, identifies the objects in given information system which might be moved to another group of objects, if values of some flexible attributes describing them will change. It identifies what is the optimal set of attributes which values should be changed to achieve the goal and what their new values should be. The support and confidence of each extended action rule is calculated. If numerical attributes will be taken into consideration, they have to be discretized first, before system DEAR is used. The effect of algorithm will be explained with the example from medical database.

Let us assume we have a decision system, with only one decision attribute, seen as *method of treatment*. Its domain contains values being integers. This decision attribute classifies objects (patients) with respect to the prognoses for patients. The cardinality of the image $d(X) = \{d_i : d(x) = d_i$ for some $x \in U\}$ is called the *rank of attribute* $\{d\}$ and is denoted by $r(d)$.

Let us observe that the decision d determines the partition $CLASS_S(d) = \{X_1, X_2, , X_{r(d)}\}$ of the set of objects X, where $X_k = d^{-1}(\{d_i\})$ for $1 \leq d_i \leq r(d)$. $CLASS_S(d)$ is called the classification of objects in S determined by the decision d.

As we mentioned before, objects correspond to patients. Also, we assume that patients in $d^{-1}(\{d_1\})$ are better prognoses for a hospital than patients in $d^{-1}(\{d_2\})$ for any $d_2 \leq d_1$. The set $d^{-1}(\{r(d)\})$ represents the patients with prognoses for complete recovery. Clearly the goal of any hospital or medical centers is to maximize the number of recovered patients. It can be achieved by shifting some patients from the group $d^{-1}(\{d_2\})$ to $d^{-1}(\{d_1\})$, for any $d_2 \leq d_1$. Namely, through special methods of treatment offered by medical centers, values of flexible attributes of some patients can be changed and the same all these patients can be moved from a group of worse prognoses ranking to a group of better prognoses.

Assume now that for any two collections of sets X, Y, we write, $X \subseteq Y$ if $(\forall x \in X)(\forall y \in Y)(x \subseteq y)$. Let $S = (X, A_{St} \cup A_{Fl} \cup \{d\})$ be a decision table and $B \subseteq A_{St} \cup A_{Fl}$. We say that attribute d depends on B if $CLASS_S(B) \subseteq CLASS_S(d)$, where $CLASS_S(B)$ is a partition of X generated by B [30].

Definition 3.8. *Assume that attribute d depends on B where $B \subseteq A_{St} \cup A_{Fl}$. The set B is called* d-reduct *in S if there is no proper subset C of B such that d depends on C.*

The concept of d-reduct in S was introduced to induce rules from S describing values of the attribute d depending on minimal subsets of $A_{St} \cup A_{Fl}$ which preserve the confidence of extracted rules. In order to induce rules in which the **then** part consists of the decision attribute d and the **if** part consists of attributes belonging to $A_{St} \cup A_{Fl}$, sub-tables $(X, B \cup \{d\})$ of S where B is a d-reduct in S should be used for rules extraction.

Table 3.2 Information system S

X	Attribute a	Attribute b	Attribute c	Attribute d
x_1	a_0	b_3	c_0	L
x_2	a_0	b_2	c_1	L
x_3	a_0	b_3	c_0	L
x_4	a_0	b_2	c_1	L
x_5	a_2	b_1	c_2	L
x_6	a_2	b_1	c_2	L
x_7	a_2	b_3	c_2	H
x_8	a_2	b_3	c_2	H

Example 3.9. Assume that $S = (\{x_1, x_2, x_3, x_4, x_5, x_6, x_7, x_8\}, \{a, c\} \cup \{b\} \cup \{d\})$ is a decision table represented by Table 3.2. The set $\{a, c\}$ lists stable attributes, $\{b\}$ is a flexible attribute and $\{d\}$ is a decision attribute. Also, we assume that H denotes patients of a good prognoses and L denotes patients of a weak prognoses.

By $L(r)$ we mean all attributes listed in the conditional part of a rule r.

If $r = [(a, a_1) * (b, b_2) \rightarrow (d, H)]$ is a rule then $L(r) = \{a, b\}$.

If $r = [[(a, a_1) * (b, b_0) * (c, c_3)] \rightarrow (d, H)]$ is a rule, then $L(r) = \{a, b, c\}$.

By $d(r)$ we denote the decision value of a rule. In our example $d(r) = H$.

If r_i, r_j are rules and $B \subseteq A_{St} \cup A_{Fl}$ is a set of attributes, then $r_i/B = r_j/B$ means that the conditional parts of rules r_i, r_j restricted to attributes B are the same.

For example, if $r_i[[(a, a_1) * (b, b_0) * (c, c_3)] \rightarrow (d, H)]$ and $r_j[[(a, a_1) * (b, b_2) * (c, c_3)] \rightarrow (d, H)]$ then $r_i/\{a, c\} = r_j/\{a, c\}$.

$CLASS_S(\{d\}) = \{\{x_1, x_2, x_3, x_4, x_5, x_6\}, \{x_7, x_8\}\}$,
$CLASS_S(\{a\}) = \{\{x_1, x_2, x_3, x_4\}, \{x_5, x_6, x_7, x_8\}\}$,
$CLASS_S(\{b\}) = \{\{x_1, x_3, x_7, x_8\}, \{x_2, x_4\}, \{x_5, x_6\}\}$,
$CLASS(\{a, b\}) = \{\{x1, x3\}, \{x2, x4\}, \{x5, x6\}, \{x7, x8\}\}$,
$CLASS_S(\{c\}) = \{\{x_1, x_3\}, \{x_2, x_4\}, \{x_5, x_6, x_7, x_8\}\}$,
$CLASS_S(\{b, c\}) = \{\{x_1, x_3\}, \{x_2, x_4\}, \{x_5, x_6\}, \{x_7, x_8\}\}$.
So, $CLASS(\{a, b\}) \subseteq CLASS_S(d)$ and $CLASS(\{b, c\}) \subseteq CLASS_S(d)$.

It can be easily checked that both $\{b, c\}$ and $\{a, b\}$ are d-reducts in S. Rules can be directly derived from d-reducts and the information system S. In our example, we get the following optimal rules:

$(a, a_0) \rightarrow (d, L)$ $\qquad\qquad$ $(c, c_0) \rightarrow (d, L)$
$(b, b_2) \rightarrow (d, L)$ $\qquad\qquad$ $(c, c_1) \rightarrow (d, L)$
$(b, b_1) \rightarrow (d, L)$ $\qquad\qquad$ $(a, a_2) * (b, b_3) \rightarrow (d, H)$
$(b, b_3) * (c, c_2) \rightarrow (d, H)$

Now, let us assume that $(a, v \rightarrow w)$ denotes the fact that the value of attribute a has been changed from v to w. Similarly, the term $(a, v \rightarrow w)(x_i)$ means that $a(x_i) = v$ has been changed to $a(x_i) = w$. Saying another words, the symptom (a, v) of a patient x_i has been changed to symptom (a, w).

Assume now that $S = (X, A_{St} \cup A_{Fl} \cup \{d\})$ is a decision table. Assume that rules r_1, r_2 have been extracted from S, B_1 is a maximal subset of A_{St} such that $r_1/B_1 = r_2/B_1$, $d(r_1) = d_1$, $d(r_2) = d_2$ and $d_1 \leq d_2$. Also, assume that $(b_1, b_2, ..., b_p)$ is a list of all attributes in $L(r_1) \cap L(r_2) \cap A_{Fl}$ on which r_1, r_2 differ and

$r_1(b_1) = v_1, r_1(b_2) = v_2, ..., r_1(b_p) = v_p$,
$r_2(b_1) = w_1, r_2(b_2) = w_2, ..., r_2(b_p) = w_p$.

By (r_1, r_2)-action rule on $x \in X$ we mean a statement:

$$((b_1, v_1 \rightarrow w_1) * (b_2, v_2 \rightarrow w_2) * ... * (b_p, v_p \rightarrow w_p))(x) \rightarrow ((d, d_1 \rightarrow d_2))(x).$$

If the value of the rule on x is true then the rule is valid. Otherwise it is false.

Let us denote by $X_{(r_1)}$ the set of all patients in X supporting the rule r_1. If (r_1, r_2)-action rule is valid on $x \in X_{(r_1)}$ then we say that the action rule supports the new profit ranking d_2 for object x.

Table 3.3 A small part of an information system S

a-(stable)	b-(flexible)	c-(stable)	e-(flexible)	g-(stable)	h-(flexible)	d-(decision)
a_1	b_1	c_1	e_1			H
a_1	b_2			g_2	h_2	L

To define an extended action rule, which formal definition will be given in Section 3.3, let us assume that two classification rules are taking into consideration. We will present them in a Table 3.3 to clarify the process of constructing an extended action rule. The classical representation of these two rules will be:

$$r_1 = (a_1 * b_1 * c_1 * e_1 \rightarrow H), r_2 = (a_1 * b_2 * g_2 * h_2 \rightarrow L).$$

Assume now that object x_i supports rule r_1 which means that it is classified as H. In order to re-classify x_i to class L, we not only need to change its value b from b_1 to b_2 but also we have to require that $g(x) = g_2$ and the value h for object x_i has to be changed to h_2. This is the meaning of the extended (r_1, r_2)-action rule given below:

$$(a_1 * (b, b_1 \rightarrow b_2) * c_1 * e_1 * (g, g_2) * (h, \rightarrow h_2))(x) \rightarrow (d, H \rightarrow L)(x).$$

The simplified version of this rule (because of stable attributes) is in form:

$$((b, b_1 \rightarrow b_2) * (g, g_2) * (h, \rightarrow h_2))(x) \rightarrow (d, H \rightarrow L)(x).$$

We assume that by $card(t)$ we mean the number of tuples having property t. By support of the extended (r_1, r_2)-action rule (given above) we mean:

$$sup(r_1, r_2) = card[(a, a_1) * (b, b_1) * (c, c_1) * (e, e_1) * (g, g_2) * (d, H)].$$

By the confidence of the extended (r_1, r_2)-action rule (given above) we mean:

$$conf(r_1, r_2) = \frac{card((a, a_1) * (b, b_1) * (c, c_1) * (e, e_1) * (g, g_2) * (d, H))}{card((a, a_1) * (b, b_1) * (c, c_1) * (e, e_1) * (g, g_2))}.$$

$$\cdot \frac{card((a, a_1) * (b, b_2) * (c, c_1) * (g, g_2) * (h, h_2) * (d, L))}{card((a, a_1) * (b, b_2) * (c, c_1) * (g, g_2) * (h, h_2))}.$$

For any extended (r_1, r_2)-action rule support and confidence can be defined in a similar way. Their formal definition is presented in Section 3.3.

Example 3.10. Assume that $S = (X, A_{St} \cup A_{Fl} \cup \{d\})$ is a decision table from the Example 3.9, $A_{Fl} = \{b\}$, $A_{St} = \{a, c\}$. It can be easily checked that rules:

$r_1 = ((b, b_1) \rightarrow (d, L))$,
$r_2 = ((a, a_2) * (b, b_3) \rightarrow (d, H))$,
$r_3 = ((b, b_3) * (c, c_2) \rightarrow (d, H))$

can be extracted from S. Clearly objects $\{x_5, x_6\} \in U_{(r1)}$.

Now, we can construct (r_1, r_2)-action rule executed on x_i:

$$((b, b_1 \rightarrow b_3))(x_i) \rightarrow ((d, L \rightarrow H))(x_i).$$

The extended (r_1, r_2)-action rule executed on x_i has the form:

$$((a, a_2) * (b, b_1 \rightarrow b_3))(x_i) \rightarrow ((d, L \rightarrow H))(x_i).$$

Clearly objects x_5, x_6 support both rules.

Algorithm to Construct Extended Action Rules

INPUT: Decision table $S = (X, A_{St} \cup A_{Fl} \cup \{d\})$, λ_1, λ_2- weights.
OUTPUT:R - set of extended action rules.
    Step 1.
Find all *d-reducts* $\{D_1, D_2, , D_m\}$ in S which satisfy the property
$\frac{card(D_i \cap A_{St})}{card(A_{St} \cup A_{Fl})} \leq \lambda_1$ (with a relatively small number of stable attributes).
    Step 2.
FOR EACH pair (D_i, D_j) of *d*-reducts (found in step 1) satisfying the property $\frac{card(D_i \cap D_j)}{card(D_i \cup D_j)} \leq \lambda_2$ **DO**
find set R_i of optimal rules in S using *d*-reduct D_i,
find set R_j of optimal rules in S using *d*-reduct D_j.
    Step 3.
FOR EACH pair of rules (r_1, r_2) in $R_i \times R_j$ having different *THEN* parts **DO**
if B_1 is a maximal subset of A_{St} such that $r_1/B_1 = r_2/B_1$, $d(r_1) = d_1$, $d(r_2) = d_2$ and $d_1 \leq d_2$, then
if $(b_1, b_2, \ldots, b_p)$ is a list of all attributes in $Dom(r_1) \cap Dom(r_2) \cap A_{Fl}$ on which r_1, r_2 differ and
$r_1(b_1) = v_1, r_1(b_2) = v_2, \ldots, r_1(b_p) = v_p$,
$r_2(b_1) = w_1, r_2(b_2) = w_2, \ldots, r_2(b_p) = w_p$
and if $(A_{St} \setminus B_1) \cap Dom(r_2) = \{a_1, a_2, \ldots, a_q\}$, and
$r_2(a_1) = u_1, r_2(a_2) = u_2, \ldots, r_2(a_q) = u_q$ and
if $[Dom(r_2) \cap A_{Fl}] \setminus \{b_1, b_2, \ldots, b_p\} = \{c_1, c_2, \ldots, c_r\}$ and $r_2(c_1) = t_1, r_2(c_2) = t_2, \ldots, r_2(c_r) = t_r$,
then the following extended (r_1, r_2)-action rule add to R:
    if $[(a_1, u_1) * (a_2, u_2) * \ldots * (a_q, u_q) * (b_1, v_1 \rightarrow w_1) * (b_2, v_2 \rightarrow w_2) * \ldots * (b_p, v_p \rightarrow w_p) * (c_1, \rightarrow t_1) * (c_2, \rightarrow t_2) * * (c_r, \rightarrow t_r)](x)$ **then** $(d, d_1 \rightarrow d_2)(x)$.
    The extended (r_1, r_2)-action rule says that if the change of values of attributes of patient x_i match the left-hand side of this rule, then the prognoses for patient x_i should change from class d_1 to class d_2.

3.2.2 System DEAR2

Action Tree algorithm is used for generating E-action rules and it is implemented as System DEAR2. The algorithm follows top-down strategy presented in [33] that searches for a solution in a part of the search space. It is
seeking at each stage for a stable attribute that has a least number of values.
Then, the set of rules is split recursively using that attribute. When all stable
attributes are processed, the final subsets, based on a decision attribute, are
split further. This method generates an action tree which is used to construct
E-action rules from the leaf nodes of the same parent. The algorithm DEAR
consists of two main steps:

1. Building Action-Tree
   a) Divide the rule table, R, taking into consideration all stable at-
      tributes.
      i. Find the domain $Dom(a)$ of each attribute $a \in A_{St}$ from the
         initial table R.
      ii. Partition the current table into sub-tables containing only rules
          supporting values of stable attributes in the corresponding sub-
          tables (assuming that the number of values in the domain of
          attribute a is the smallest).
   b) Divide each lowest level sub-table into new sub-tables containing rules
      with the same decision value.
   c) Represent each leaf as a set of rules which do not contradict on stable
      attributes and also define decision value d_i. The path from the root
      to that leaf gives the description of objects supported by these rules.
2. Generating action rules
   a) Comparing all unmarked leaf nodes of the same parent form action
      rules.
   b) Calculate the support and the confidence for each rule. If both of
      them are above minimal thresholds, the rule is extracted, and added
      to the knowledgebase.

The algorithm starts with all extracted classification rules at the root node
of the tree. We select one stable attribute from the set of all stable attributes
to partition the rules. For each value of the attribute a branch is created, and
the corresponding subset of rules, that have the attribute value specified by
the branch, is moved to the newly created child node. This step is repeated
for each node.

When all stable attributes are taken into consideration, the last split is
based on a decision attribute for each branch. The node selection is based on
the stable attributes with the smallest number of possible values among all
the remaining stable attributes.

An action tree has two types of nodes: a leaf node and a non-leaf one. At
a non-leaf node in the tree, the set of rules is partitioned along the branches
and each child node gets its corresponding subset of rules. When the stable

attributes have the same value, every path to the decision attribute node, one level above the leaf node, represents a subset of the extracted classification rules. Each leaf represents a set of rules, which define decision value d_i and do not contradict on stable attributes. The path from the root to that leaf gives the description of objects supported by these rules.

Table 3.4 Decision System S

X	Attribute a	Attribute b	Attribute c	Attribute d
x_1	a_1	b_2	c_1	H
x_2	a_1	b_2	c_1	H
x_3	a_2	b_1	c_1	A
x_4	a_2	b_1	c_1	A
x_5	a_1	b_2	c_2	A
x_6	a_1	b_2	c_2	A
x_7	a_1	b_0	c_1	H
x_8	a_1	b_0	c_1	H
x_9	a_1	b_0	c_3	H
x_{10}	a_1	b_1	c_2	H
x_{11}	a_2	b_2	c_3	A
x_{12}	a_2	b_0	c_1	A

Example 3.11. Let us take Table 3.4 as an example of the decision system S. We assume that $A_{St} = \{a, b\}$, and $A_{Fl} = \{c, d\}$. The goal is to reclassify some objects from class (d, H) into class (d, A). First we create the knowledgebase, with the set R of certain rules from decision system S, in a form of another table (Table 3.5). The first column of this table shows objects in S supporting the rules from R (each row represents one rule). For instance, the first row represents a rule $(a, a_2) \rightarrow (d, A)$, second row represents a rule $[(a, a_1) * (c, c_1)] \rightarrow (d, H)$, third row represents a rule $[(a, a_1) * (b, b_0)] \rightarrow (d, H)$, etc.

To construct an action tree first we start with the set R as a table **T1** at the root of the tree. The root node selection is based on a stable attribute

Table 3.5 Set of rules with objects corresponding to them (**T1**)

Set of objects	Attribute a	Attribute b	Attribute c	Attribute d
$\{x_3, x_4, x_{11}, x_{12}\}$	a_2			A
$\{x_1, x_2, x_7, x_8\}$	a_1		c_1	H
$\{x_7, x_8, x_9\}$	a_1	b_0		H
$\{x_3, x_4\}$		b_1	c_1	A
$\{x_5, x_6\}$		b_2	c_2	A

with the smallest number of states among all stable attributes. We repeat
this for all stable attributes. Next the tree is split based on the value of the
decision attribute. In our example in **T1**, we use attribute a to split the

T2 Set of objects	a	b	c	d
$\{x_1, x_2, x_7, x_8\}$	a_1		c_1	H
$\{x_7, x_8, x_9\}$	a_1	b_0		H
$\{x_3, x_4\}$		b_1	c_1	A
$\{x_5, x_6\}$		b_2	c_2	A

T3 Set of objects	a	b	c	d
$\{x_3, x_4, x_{11}, x_{12}\}$	a_2			A
$\{x_3, x_4\}$		b_1	c_1	A
$\{x_5, x_6\}$		b_2	c_2	A

table into two sub-tables **T2** and **T3** defined by values $\{a_1, a_2\}$ of attribute
a respectively. It follows from the fact that $card[V_a] < card[V_b]$. In such way
we form two tables: one (**T2**) with rules containing $a = a_1$ and the second
(**T3**) with rules containing $a = a_2$. We can notice that all objects in the sub-
table **T3** have the same decision attribute (d, A), therefore no action rules
can be generate from this sub-table. So this table will no longer be taken into
consideration and division in terms of other attributes. Because sub-table **T2**
contains different decision values and there is another stable attribute b, it
is divided into next sub-tables **T4**, **T5**, and **T6** corresponding to the values
of attribute b: **T4** for value b_0, **T5** for value b_1 and **T6** for value b_2. At this
point, each sub-table does not contain any stable attributes.

T4 Set of objects	a	b	c	d
$\{x_1, x_2, x_7, x_8\}$	a_1		c_1	H
$\{x_7, x_8, x_9\}$	a_1	b_0		H

T5 Set of objects	a	b	c	d
$\{x_1, x_2, x_7, x_8\}$	a_1		c_1	H
$\{x_3, x_4\}$		b_1	c_1	A

T6 Set of objects	a	b	c	d
$\{x_1, x_2, x_7, x_8\}$	a_1		c_1	H
$\{x_5, x_6\}$		b_2	c_2	A

Table **T4** cannot be divided any further (as it was in **T3** case). In sub-table **T5** there is only one value of flexible attribute $c = c_1$, so this table also cannot be partitioned. The remaining table **T6** is partitioned into two independent sub-tables **T7** and **T8** depending on values of decision attribute d: **T7** for value $d = H$ and **T8** for value $d = A$.

T7 Set of objects	a	b	c	d
$\{x_1, x_2, x_7, x_8\}$	a_1		c_1	H

T8 Set of objects	a	b	c	d
$\{x_1, x_2, x_7, x_8\}$	a_1		c_1	H

The path from the root to the leaf described by d gives the information of rules and objects supported by them.

For example following the path labeled by $[a = a_1]$, $[b = b_2]$ and $[d = H]$ we get sub-table **T7**, while using path $[a = a_1]$, $[b = b_2]$ and $[d = A]$ we get **T8**. We can compare pairs of rules belonging to these two tables and construct the action rule such as:

$$r = [[(a, a_1) * (c, c_1 \rightarrow c_2)] \rightarrow (d, H \rightarrow A)].$$

We evaluate the rule by checking its support and confidence. they are equal: $sup(r) = min\{4, 2\} = 2$, $conf(r) = 1 \cdot \frac{2}{3} = \frac{2}{3}$.

The action-tree algorithm proposed here requires the extraction of all classification rules from the decision system before any action rule is constructed and has $0(k^2)$ complexity in the worst case, where k is the number of classification rules. This new algorithm was implemented and tested on several data sets from UCI Machine Learning Repository and was more efficient then other algorithms [60], [61], [62], [63], [64].

3.3 E-action Rules

Let $S = (X, A_{St} \cup A_{Fl} \cup \{d\})$ be a decision system, where $d \notin A_{St} \cup A_{Fl}$. To improve the efficiency of the algorithm, when the number of attributes is large, we can extract rules from sub-tables $(X, B \cup \{d\}) \subseteq S$, where B is a $d - reduct$ of system S [30].

Let us assume that r_i, r_j are rules extracted from S. The idea of extended action rule (e-action rule) was given in [67] and extended in [49]. Its definition is given below.

We assume that:

1. $B_{St} \subseteq A_{St}$ is maximal, such that $r_i/B_{St} = r_j/B_{St}$
2. $d(r_i) = d_i, d(r_j) = d_j$ and $d_i \leq d_j$ (d_i is ranked lower than d_j)
3. $(\forall a \in [A_{St} \cap L(r_i) \cap L(r_j)])[a(r_i) = a(r_j)]$
4. $(\forall m \leq q)(\forall e_m \in [A_{St} \cap (L(r_j) \setminus L(r_i))])[e_m(r_j) = u_j]$
5. $(\forall m \leq r)(\forall c_m \in [A_{Fl} \cap (L(r_j) \setminus L(r_i))])[c_m(r_j) = t_j]$
6. $(\forall m \in p))(\forall b_m \in [A_{Fl} \cap L(r_i) \cap L(r_j)])([b_m(r_i) = v_m] * [b_m(r_j) = w_m])$

Assume now that rules r_i, r_j are extracted from S and $r_i/A_{St} = r_j/A_{St}$, $d(r_i) = d_i$, $d(r_j) = d_j$ and $d_i \leq d_j$. We also have the assumption that $(b_1, b_2, ..., b_m)$ is a list of all attributes in $Dom(r_i) \cap Dom(r_j) \cap A_{Fl}$ on which r_i and r_j differ and $r_i(b_1) = v_1$, $r_i(b_2) = v_2$, ... , $r_i(b_m) = v_m$, $r_j(b_1) = w_1$, $r_j(b_2) = w_2$, ... , $r_j(b_m) = w_m$.

Definition 3.12. *By (r_i, r_j)-action rule on $x \in X$ we mean an expression of the form:*
$$r = [(b_1, v_1 \rightarrow w_1) * (b_2, v_2 \rightarrow w_2) * ... * (b_m, v_m \rightarrow w_m)](x) \rightarrow (d, d_i \rightarrow d_j)(x).$$

Object $x \in X$ supports r_i, r_j e-action rule r in system $S = (X, A_{St} \cup A_{Fl} \cup \{d\})$ if the following conditions are satisfied:

1. $(\forall i \leq p)$ $[b_i \in L(r)][b_i(x) = v_i] \wedge d(x) = d_1$
2. $(\forall i \leq p)$ $[b_i \in L(r)][b_i(y) = w_i] \wedge d(y) = d_2$
3. $(\forall j \leq p)$ $[a_j \in A_{St} \cap L(r_j)][a_j(x) = u_j]$
4. $(\forall j \leq p)$ $[a_j \in A_{St} \cap L(r_j)][a_j(y) = u_j]$
5. Objects x, y support rules r_1, r_2 respectively

Definition 3.13. *By the support of rule r we mean the number of all objects in S satisfying the left side of the rule consisting of conditional parts of terms.*

$$sup(r) = card[(b_1, v_1) * (b_2, v_2) * ... * (b_m, v_m) * (d, d_i)].$$

For computing the confidence of extended (r_1, r_2)-*action rule* we divide the number of objects supporting (r_1, r_2)-*action rule* by the number of objects supporting left hand side of this rule and multiply it by the confidence of the classification rule r_2. Values of stable attributes listed in r_1 do not have to be considered at all.

Definition 3.14. *The confidence of extended (r_1, r_2)-action rule is equal to:*

$$conf((r_1, r_2)) = \frac{sup(r)}{sup(L(r))} \cdot conf(r_2).$$

Example 3.15. Let us assume we have extended (r_1, r_2)-action rule r in form:

$$r = [(a, a_1) * (c, c_1 \to c_2) \to (d, H \to A)].$$

The support of this rule is $sup(r) = 4$, while the confidence is equal $conf(r) = \frac{4}{4} \cdot \frac{2}{3} = \frac{2}{3} = 66\%$.

In more general scenario, let us assume that $S = (X, A_{St} \cup A_{Fl} \cup \{d\})$ is a decision system. We say that x is a d_1-object if $d(x) = x_1$. We also assume that $(a_1, a_2, ..., a_m) \subseteq A_{St}$, $(b_1, b_2, ..., b_n) \subseteq A_{Fl}$ and $a_{i,j}, b_{i,j}$ denote rules of attributes a_i and b_i respectively, and that

$$r = [(a_{1,1} * a_{2,1} * ... * b_{1,1} * b_{2,1} * ... * b_{n,1}) \to d_1]$$

is a classification rule extracted from S supporting some d_1-objects in S. Our goal is again to re-classify objects from class d_2 to d_1 class.

By an action rule $r_{[d_2 \to d_1]}$ associated with r we mean the expression of the form:

$$r_{[d_2 \to d_1]} = [[a_{1,1} * a_{2,1} * ... * (b_1, \to b_{1,1}) * (b_2, \to b_{2,1}) * ... * (b_n, \to b_{n,1})] \to (d, d_2 \to d_1)].$$

The support of the action rule $r_{[d_2 \to d_1]}$ is defined as:

$$sup(r_{[d_2 \to d_1]}) = card\{x \in X : (a_1(x) = a_{1,1}) * (a_2(x) = a_{2,1}) * ... * (a_m(x) = a_{m,1}) * (d(x) = d_2)\}.$$

If $sup(r_{[d_2 \to d_1]}) = 0$, then rule $r_{[d_2 \to d_1]}$ is not interesting and cannot be used for reclassification of objects.

Let $sup(r_{[\to d_1]}) = \cup\{sup(r_{[d_2 \to d_1]}) : (d_2 \neq d_1) \wedge d_2 \in V_d\}$ and $sup(R_{[\to d_1]}) = \cup\{sup(r_{[\to d_1]}) : r \in R(d_1)\}$, where $R(d_1)$ is the set of all classification rules defining d_1. So $sup(R_S) = \cup\{sup(r_{[\to d_1]}) : d_1 \in V(d)\}$ contains all objects in S which potentially can be reclassified.

Assume now that $X(d_1) = \{x \in X : d(x) \neq d_1\}$. Objects in the set $B(d_1) = X(d_1) - sup(R_{[\to d_1]})$ can not be reclassified to the class d_1 and they are called d_1-*resistant*.

The set $B(\neg d_1) = \cap\{B(d_i) : (d_i \neq d_1) \wedge (d_i \in V_d)\}$ represents the set of d_1-objects which can not be reclassified. They are called d_1-*stable*. Similarly, the set $B_d = \cup\{B(\neg d_i) : d_i \in V_d\}$ represents objects in X which can not be reclassified to any decision class. All these objects are called d-*stable*.

The question is, how to find these objects?

Let $r_{i[d_2 \to d_1]}$ and $r_{j[d_2 \to d_3]}$ are two action rules extracted from S.

We say that they are equivalent $(\cong)$, $r_{i[d_2 \to d_1]} \cong r_{j[d_2 \to d_3]}$ if $\forall (b_i \in A_{St} \cup A_{Fl})[r_i/b_i = r_j/b_i]$. Two rules belong to the same equivalence class if values of their stable attributes are not conflicting each other.

Let us take d_2-object $x \in sup(r_{i[d_2 \to d_1]})$. We say that x positively supports $r_{i[d_2 \to d_1]}$ if there is no classification rule r_j extracted from S and describing $d_3 \in V_d, d_3 \neq d_1$, equivalent to r_i, such that $x \in sup(r_{j[d_2 \to d_3]})$. The corresponding subset of support is denoted by $sup^+(r_{i[d_2 \to d_1]})$. Otherwise, we say that object x negatively supports $r_{i[d_2 \to d_1]}$, and in such case the subset of support is denoted by $sup^-(r_{i[d_2 \to d_1]})$. By the confidence of rule $r_{i[d_2 \to d_1]}$ we mean

$$conf(r_{i[d_2 \to d_1]}) = \frac{card[sup^+(r_{i[d_2 \to d_1]})]}{card[sup(r_{i[d_2 \to d_1]})]} \cdot conf(r).$$

We can observe that a set $B^+(d_1) = [X(d_1) - sup^+(R_{[\to d_1]})]$ contains all d_1-resistant objects. The set $B^+(\neg d_1) = \cap\{B^+(d_i) : (d_i \neq d_1) \wedge (d_i \in V_d)\}$ is the set of all d_1-stable objects in S. The same $B_d^+ = \cap\{B_d^+ = \cup B^+(\neg)d_i\} :$ $(d_i \in V_d)\}$ contains all d-stable objects in S.

Moving Objects between Classes

For objects in system S we can assign class-movability index which is helpful in dividing the set of objects into different rank groups. Objects of higher rank are seen as objects more interesting to move between decision classes than objects of lower rank.

Let $(V_d, \preceq)$ be an ordered set of values of the decision attribute d, where $V = \{d_1, d_2, ..., d_k\}$. By function F_S, called *decision attribute ranking* associated with S we mean assignment to each $d_i, 1 \leq i \leq k$, an integer number.

If $F_S(d_i) \leq F_S(d_j)$, then $d_i \preceq d_j$.

For each $j \in \{1, 2, ..., k\}$ we start with a collection of sets $P_j(i) = sup^+(r_{[d_j \to d_i]})$, each containing all positively supported d_j-objects in S. It means $P_j(i)$ contains all objects in X which can be reclassified (moved) from one decision class d_j to another d_i. $P_j(N) = \cap\{\{P_j(i)\} : i \in N\}$(for any $N \subseteq \{1, 2, ..., k\}$) contains all objects in X which can be moved from one decision class d_j to any of the classes d_i, where $i \in N$.

The set $N_j^+ = \{i \in N : F_S(d_i) - F_S(d_j) > 0\}$, for any $N \subseteq \{1, 2, ..., k\}$, represents all attribute values in $\{d_i : i \in N\}$ which have higher decision attribute ranking than d_j.

By class-movability index assigned to N_j, we mean:

$$ind(N_j) = \sum \{F_S(d_i) - F_S(d_j) : i \in N_j^+\}.$$

By class-movability index assigned to d_j-object x, we mean:

$$ind_S(x) = max\{ind(N_j) : N_j \subseteq \{1, 2, ..., k\} \wedge x \in P_j(N)\}.$$

Index $ind_S(x)$ shows the overall amount of improvement open to object x in terms of a number of available re-classifications and the improvements linked with every re-classification of the object.

Example 3.16. Let us assume an information system $S = (X, A_{St} \cup A_{Fl} \cup \{d\})$ is represented as Table 3.6, where $X = \{x_i\}_{i \in \{1,2,...,8\}}$, $A_{St} = \{a, b, c\}$, $A_{Fl} = \{e, f, g\}$. We are interested in reclassifying d_2-objects either to class d_1 or d_3. In order to do this, first we extract certain rules from this system S using one of the rules discovery methods, e.g. LERS. We received the following classification rules:

Table 3.6 Information system S with decision class d

X	Attr. a	Attr. b	Attr. c	Attr. e	Attr. f	Attr. g	Attr. d
x_1	a_1	b_1	c_1	e_1	f_1	g_2	d_3
x_2	a_2	b_1	c_2	e_2	f_2	g_2	d_1
x_3	a_1	b_1	c_3	e_2	f_3	g_2	d_2
x_4	a_2	b_1	c_1	e_2	f_1	g_2	d_2
x_5	a_1	b_2	c_1	e_3	f_1	g_2	d_2
x_6	a_1	b_1	c_2	e_2	f_1	g_3	d_2
x_7	a_2	b_3	c_2	e_2	f_2	g_2	d_2
x_8	a_1	b_1	c_2	e_3	f_3	g_2	d_2

$r_1 : (a_1 * b_1 * f_1 * g_2) \to d_3$
$r_2 : (b_1 * c_2 * e_2 * g_2) \to d_1$
$r_3 : e_1 \to d_3$
$r_4 : (b_1 * f_2) \to d_1$.

We can notice that $R_{[d,\to d_1]} = \{r_2, r_4\}$ and $R_{[d,\to d_3]} = \{r_1, r_3\}$. Action rules associated with these four classification rules are:

$r_{1[d_2 \to d_3]} : [a_1 * b_1 * (f, \to f_2) * (g, \to g_1)] \to (d, d_2 \to d_3)$
$r_{2[d_2 \to d_1]} : [b_1 * c_2 * (e, \to e_2) * (g, \to g_2)] \to (d, d_2 \to d_1)$
$r_{3[d_2 \to d_3]} : (e, \to e_1) \to (d, d_2 \to d_3)$
$r_{4[d_2 \to d_1]} : [b_1 * (f, \to f_2)] \to (d, d_2 \to d_1)$.

It can be easily shown that

$sup(r_{1[d_2 \to d_3]}) = \{x_3, x_6, x_8\}$
$sup(r_{2[d_2 \to d_1]}) = \{x_6, x_8\}$
$sup(r_{3[d_2 \to d_3]}) = \{x_3, x_4, x_5, x_6, x_7, x_8\}$
$sup(r_{4[d_2 \to d_1]}) = \{x_3, x_4, x_6, x_8\}$.

So $sup(r_{3[\to d_3]}) = \{x_3, x_4, x_5, x_6, x_7, x_8\}$, $sup(r_{4[\to d_1]}) = \{x_3, x_4, x_6, x_8\}$. All d_2-objects can be potentially reclassified to the class d_1.

On the other hand, d_2-objects only from the set $\{x_3, x_4, x_6, x_8\}$ can be potentially reclassified to the same class d_1. It means that d_2-objects $\{x_5, x_7\}$ are d_1-resistant.

Let us also notice that action rules $r_{1[d_2 \to d_3]}$ and $r_{2[d_2 \to d_1]}$ are equivalent. $sup(r_{1[d_2 \to d_3]}) = \{x_3, x_6, x_8\}$ and $sup(r_{2[d_2 \to d_1]}) = \{x_6, x_8\}$ then only x_3 positively supports $r_{1[d_2 \to d_3]}$ and objects in $\{x_6, x_8\}$ negatively support both $r_{1[d_2 \to d_3]}$ and $r_{2[d_2 \to d_1]}$.

The confidence:

$$conf(r_{1[d_2 \to d_3]}) = \frac{card[sup^+(r_{1[d_2 \to d_3]})]}{card[sup(r_{1[d_2 \to d_3]})]} \cdot conf(r_1) = \frac{1}{3} \cdot 1 = \frac{1}{3}.$$

$$conf(r_{2[d_2 \to d_1]}) = \frac{card[sup^+(r_{2[d_2 \to d_1]})]}{card[sup(r_{2[d_2 \to d_1]})]} \cdot conf(r_2) = \frac{0}{2} \cdot 1 = 0.$$

3.3.1 ARAS Algorithm

The process of constructing action rules from pairs of classification rules is not only sometimes unnecessarily expensive but also gives too much possibilities in constructing their classification parts. In [40] it was shown that action rules do not have to be built from pairs of classification rules and that single classification rules are also sufficient to achieve the same goal. We propose a very simple LERS-type algorithm for constructing action rules from a single classification rule. LERS is a classical example of a bottom-up strategy which constructs rules with a conditional part of the length $k+1$ after all rules with a conditional part of the length k have been constructed. Relations representing rules produced by LERS are marked. System ARAS assumes that LERS is used to extract classification rules. This way ARAS instead of verifying the validity of certain relations has to check if these relations are marked earlier by LERS. The same, by using LERS as the pre-processing module for ARAS, the overall complexity of the algorithm is decreased. A bottom-up strategy, generating action rules from single classification rules was proposed in [53]. We explain this method on the examples.

Table 3.7 Incomplete Information system S with decision class d (**T1**)

X	Attribue a	Attribute b	Attribute c	Attribute e	Attribute g	Attribute d
x_1	a_2			e_1	g_4	I
x_2	a_2	b_3	H	e_2	g_4	I
x_3			H	e_4	g_1	A
x_4	a_1	b_2	L	e_3	g_3	A
x_5	a_2	b_0	L		g_5	E
x_6	a_2	b_0	H	e_1	g_2	E
x_7		b_0	H	e_1		A
x_8	a_2			e_2	g_4	E
x_9	a_3	b_0	H	e_2	g_4	A
x_{10}	a_3	b_1	L	e_1	g_1	A
x_{11}	a_3	b_1	L	e_2	g_0	A
x_{12}	a_2	b_3	H	e_4	g_1	I

Example 3.17. Assume that $S = (X, A_{St} \cup A_{Fl} \cup \{d\})$ is decision system, where $A_{St} = \{a, b\}, A_{Fl} = \{c, e, g\}$, and $\{d\}$ is the decision. For a generality reason, we take an incomplete decision system. It is represented as Table 3.7. We start with the incomplete decision system S as the root of the Reduction Tree (**T1**). If we want to reclassify objects from class(d, A) into class (d, E) or (d, I), we have to split S into sub-tables first, taking an attribute with the minimal number of distinct values as the splitting one. In our example, we chose attribute a. In such way we obtain three sub-tables: **T2** for value a_1, **T3** for value a_2 and **T4** for value a_3. Because objects x_3 and x_7 in S contain null

T2 X	b	c	e	g	d
x_3		H	e_4	g_1	A
x_4	b_2	L	e_3	g_3	A
x_7	b_0	H	e_1		A

T3 X	b	c	e	g	d
x_1			e_1	g_4	I
x_2	b_3	H	e_2	g_4	I
x_3		H	e_4	g_1	A
x_5	b_0	L		g_5	E
x_6	b_0	H	e_1	g_2	E
x_7	b_0	H	e_1		A
x_8			e_2	g_4	E
x_{12}	b_3	H	e_4	g_1	I

T4 X	b	c	e	g	d
x_3		H	e_4	g_1	A
x_7	b_0	H	e_1		A
x_9	b_0	H	e_2	g_4	A
x_{10}	b_1	L	e_1	g_1	A
x_{11}	b_1	L	e_2	g_0	A

values in column a, we move them both to all three newly created sub-tables. We will follow optimistic approach in the process of action rules discovery, which means that the null value is interpreted as the disjunction of all possible attribute values in the corresponding domain. This process is then recursively continued for all stable attributes. Sub-tables corresponding to outgoing edges from the root node which are labeled by **T2**, **T4** are removed because they do not contain decision value (d, I). Any remaining node in the resulting tree can be used for discovering action rules. Clearly, if node n is used to construct action rules, then its children are not used for that purpose.

Next example will describe ARAS algorithm in details.

Example 3.18. Let us assume that the decision system $S = (X, A_{St} \cup A_{Fl} \cup \{d\})$, is represented by Table 3.8. A number of different methods can be used to extract rules in which the **then** part consists of the decision attribute $\{d\}$ and the conditional part consists of attributes belonging to $A_{St} \cup A_{Fl}$. In our example we take into consideration complete information system, but for the incomplete one, the method is similar. The set $A_{St} = \{a, b, c\}$ contains stable attributes and $A_{Fl} = \{e, f, g\}$ contains flexible attributes. System LERS [13] is used to extract classification rules. We are interested in reclassifying (d, A)-objects either to class $(d, I) = d_I$ or $(d, E) = d_E$. Four certain

Table 3.8 Information system S with decision class d

X	Attr. a	Attr. b	Attr. c	Attr. e	Attr. f	Attr. g	Decision d
x_1	a_1	b_0	H	e_1	f_1	g_0	I
x_2	a_2	b_0	H	e_2	f_1	g_2	A
x_3	a_3	b_0	H	e_3	f_1	g_2	A
x_4	a_1	b_0	L	e_3	f_1	g_0	A
x_5	a_1	b_1	H	e_2	f_1	g_0	A
x_6	a_2	b_0	H	e_3	f_2	g_0	A
x_7	a_2	b_2	L	e_3	f_1	g_1	A
x_8	a_2	b_0	L	e_3	f_1	g_1	E

classification rules describing either (d, I) or (d, E) are discovered by LERS from the decision system S. They are given below:

$r_1 = [(b_0 * c_H * f_1 * g_0) \rightarrow d_I]$
$r_2 = [(a_2 * b_0 * e_3 * f_1) \rightarrow d_E]$
$r_3 = [e_1 \rightarrow d_I]$
$r_4 = [(b_0 * g_1) \rightarrow d_E]$.

Action rule schemas associated with r_1, r_2, r_3, r_4 and the reclassification task either $[d, (d_A \rightarrow d_I)]$ or $[d, (d_A \rightarrow d_E)]$ are:

$r_{1[(d_A \rightarrow d_I)]} = [b_0 * c_H * (f, \rightarrow f_1) * (g, \rightarrow g_0)] \rightarrow (d, d_A \rightarrow d_I)$
$r_{2[(d_A \rightarrow d_E)]} = [a_2 * b_0 * (e, \rightarrow e_3) * (f, \rightarrow f_1)] \rightarrow (d, d_A \rightarrow d_E)$
$r_{3[(d_A \rightarrow d_I)]} = (e, \rightarrow e_1) \rightarrow (d, d_A \rightarrow d_I)$
$r_{4[(d_A \rightarrow d_E)]} = [b_0 * (g, \rightarrow g_1)] \rightarrow (d, d_A \rightarrow d_E)$.

We can show that:

$Sup(r_{1[(d_A \rightarrow d_I)]}) = \{x_2, x_3, x_6\}$,
$Sup(r_{2[(d_A \rightarrow d_E)]}) = \{x_2, x_6\}$,
$Sup(r_{3[(d_A \rightarrow d_I)]}) = \{x_2, x_3, x_4, x_5, x_6, x_7\}$,
$Sup(r_{4[(d_A \rightarrow d_E)]}) = \{x_2, x_3, x_4, x_6\}$.

Assuming that:

$X(r_1, d_A) = Sup(r_{1[(d_A \rightarrow d_I)]})$,
$X(r_2, d_A) = Sup(r_{2[(d_A \rightarrow d_E)]})$,
$X(r_3, d_A) = Sup(r_{3[(d_A \rightarrow d_I)]})$,
$X(r_4, d_A) = Sup(r_{4[(d_A \rightarrow d_E)]})$

and by applying ARAS algorithm we get:

$(b_0 * c_H * a_1)^* = \{x_1\} \nsubseteq X(r_1, d_A)$,
$(b_0 * c_H * a_2)^* = \{x_2, x_6\} \subseteq X(r_1, d_A)$
$(b_0 * c_H * f_2)^* = \{x_6\} \subseteq X(r_1, d_A)$
$(b_0 * c_H * g_1)^* = \{x_7, x_8\} \nsubseteq X(r_1, d_A)$
$(b_0 * c_H * g_2)^* = \{x_2, x_3\} \subseteq X(r_1, d_A)$.

Algorithm ARAS will construct two action rules for the first action rule schema:

$$[b_0 * c_H * (f, f_2 \rightarrow f_1) * (g, \rightarrow g_0)] \rightarrow (d, d_A \rightarrow d_I)$$

$$[b_0 * c_H * (f, \rightarrow f_1) * (g, g_2 \rightarrow g_0)] \rightarrow (d, d_A \rightarrow d_I)$$

In a similar way we construct action rules from the remaining three action rule schemas. Algorithm ARAS consists of two main modules. To explain them in a better way, we use another example which has no connection with previous tables. The first module of ARAS extracts all classification rules from S following LERS strategy.

Example 3.19. Assuming that $\{d\}$ is the decision attribute and user is interested in reclassifying objects from its value (d, A) to (d, H), we treat the rules defining (d, A) as seeds and build clusters around them.

For instance, if $A_{St} = \{a, b, g\}$ and $A_{Fl} = \{c, e, h\}$ are attributes in $S = (X, A_{St} \cup A_{Fl} \cup \{d\})$, and $r = [(a_1 * b_1 * c_1 * e_1) \rightarrow d_1]$ is a classification rule in S, where $V_a = \{a_1, a_2, a_3\}$, $V_b = \{b_1, b_2, b_3\}$, $V_c = \{c_1, c_2, c_3\}$, $V_e = \{e_1, e_2, e_3\}$, $V_g = \{g_1, g_2, g_3\}$, $V_h = \{h_1, h_2, h_3\}$, then we remove from S all tuples containing values $a_2, a_3, b_2, b_3, c_1, e_1$ and we use again LERS to extract rules from the obtained subsystem. Each rule defining (d, H) is used jointly with r to construct an action rule. The validation step of each of the set-inclusion relations, in the second module of ARAS, is replaced by checking if the corresponding term was marked by LERS in the first module of ARAS.

The complexity of ARAS is lower than the complexity of DEAR system discovering action rules. The justification here is quite simple. DEAR system [66] groups classification rules into clusters of non-conflicting rules and then takes all possible pairs of classification rules within each cluster and tries to build action rules from them. ARAS algorithm is treating each classification rule describing the target decision value as a seed and grabs other classification rules describing non-target decision values to form a cluster. Then it builds decision rules automatically from them. Rules grabbed into a seed are only compared with that seed. So, the number of pairs of rules which have to be checked, in comparison to DEAR is greatly reduced. So, the complexity of the second module of ARAS is $0(k \cdot n)$, where n is the number of classification rules extracted by LERS and k is the number of clusters. In comparison time complexity of the second module of DEAR is equal to $0(n \cdot n)$, where n is the same as in ARAS.

3.4 Action Rules Tightly Coupled Framework

The method presented below generates a complete set of shortest action rules without using pre-existing classification rules. During many experiments authors noticed that the flexibility of attributes are not equal. E.g. health conditions are more flexible than the social conditions. In paper [34] we can find first steps taken with problem of mining action rules without pre-existing classification rules. The proposed algorithm was similar to Apriori [3]. The definition of an action rule in [52] allows changes on stable attributes. Changing the value of an attribute, either stable or flexible, is linked with a cost [48], [61]. In order to rule out action rules with undesired changes on stable attributes, authors assigned very high cost to such changes. That way, the cost of action rules discovery is getting unnecessarily increased. The dependencies between attribute values which are linked in a natural way with the cost of rules used to accept or reject the rule were not taken into consideration. The method for discovering action rules directly from the decision system S is also based (like previous algorithms) on Pawlak's model of an information system. In [20] authors proposed a method that extracts action rules directly from attribute values in incomplete information systems without using pre-existing conditional rules. This means that they use pre-existing classification rules or generated them using rule discovery algorithms such as LERS or ERID [4], [5], then construct action rules either from certain pairs of the rules or from a single classification rule. The methods in [16], [40], [53] do not formulate actions directly from existing classification rules. Actions are built as the effect of possible changes in different classification rules. Thus, the extraction of classification rule during action rule formulation is inevitable.

The new algorithm ARD for constructing action rules is similar to ERID. Its main goal is to identify certain relationships between granules defined by the indiscernibility relation on system's objects. Some of these relationships uniquely define action rules for S. Papers [6], [40], [42] present a new strategy for discovering action rules directly from the decision system. To present this method, it is sufficient to show how terms of length greater than one are built. Only positive marks yield action rules. Action terms of length k are built from unmarked action terms of length $k-1$ and unmarked atomic action terms of length one.

Let us assume, that $S = (X, A \cup \{d\}, V)$ is a decision system, and λ_1, λ_2 denote minimum support and minimum confidence respectively. Each attribute $a \in A$ defines in a unique way the set $C_S(a) = \{N_S(t_a) : t_a$ is an atomic action term built from elements in $V_a\}$.

Marking Strategy:
$\forall N_S(t_a) \in C_S(a)$:

- If $L(N_S(t_a)) = 0$ or $R(N_S(t_a)) = 0$ or $L(N_S(t_a * t_d)) = 0$ or $R(N_S(t_a * t_d)) = 0$ then t_a is marked negative
- If $L(N_S(t_a)) = R(N_S(t_a))$ then t_a stays unmarked
- If $card(L(N_S(t_a * t_d))) < \lambda_1$ then t_a is marked negative
- If $card(L(N_S(t_a * t_d))) \geq \lambda_1$ and $conf(t_a \rightarrow t_d) < \lambda_2$ then t_a stays unmarked
- If $card(L(N_S(t_a * t_d))) \geq \lambda_1$ and $conf(t_a \rightarrow t_d) \geq \lambda_2$ then t_a is marked positive

From all marked forms, action rule $t_a \rightarrow t_d$ is taken into consideration.

Example 3.20. Let us take once again decision system S from Table 2.14 with $A_{St} = \{a, c\}$, $A_{Fl} = \{b, d\}$. We are interested in objects re-classification from decision class (d, H) to (d, A). Assume, that threshold for minimal support is $\lambda_1 = 2$ and for minimal confidence $\lambda_2 = 0.25$.

First Loop:
Building all atomic action terms for S.

For decision attribute $\{d\}$ in S:

$N_S(t_{11}) = [\{x_1, x_2, x_6, x_9, x_{10}\}, \{x_3, x_4, x_5, x_7, x_8\}]$

For classification attributes, both stable and flexible, in S:

$$t_2 = (a, a_2 \rightarrow a_2) \qquad t_3 = (a, a_3 \rightarrow a_3) \qquad t_4 = (c, c_1 \rightarrow c_1)$$
$$t_5 = (c, c_2 \rightarrow c_2) \qquad t_6 = (c, c_3 \rightarrow c_3) \qquad t_7 = (b, b_1 \rightarrow b_1)$$
$$t_8 = (b, b_1 \rightarrow b_2) \qquad t_9 = (b, b_2 \rightarrow b_1) \qquad t_{10} = (b, b_2 \rightarrow b_2)$$

$N_S(t_1) = [\{x_1, x_8, x_9\}, \{x_1, x_8, x_9\}]$	not marked as $Y_1 = Y_2$
$N_S(t_2) = [\{x_2, x_3, x_5, x_6, x_{10}\}, \{x_2, x_3, x_5, x_6, x_{10}\}]$	not marked as $Y_1 = Y_2$
$N_S(t_3) = [\{x_4, x_7\}, \{x_4, x_7\}]$	*marked negative* as $card(Y_1 \cap Z_1) = 0$
$N_S(t_4) = [\{x_1, x_8\}, \{x_1, x_8\}]$	not marked as $Y_1 = Y_2$
$N_S(t_5) = [\{x_4, x_5, x_6, x_7\}, \{x_4, x_5, x_6, x_7\}]$	not marked as $Y_1 = Y_2$
$N_S(t_6) = [\{x_2, x_3, x_9, x_{10}\}, \{x_2, x_3, x_9, x_{10}\}]$	not marked as $Y_1 = Y_2$
$N_S(t_7) = [\{x_1, x_3, x_4, x_5, x_7, x_9\}, \{x_1, x_3, x_4, x_5, x_7, x_9\}]$	not marked as $Y_1 = Y_2$
$N_S(t_8) = [\{x_1, x_3, x_4, x_5, x_7, x_9\}, \{x_2, x_6, x_8, x_{10}\}]$	not marked as $sup = 2$ but $conf = 0.04 < \lambda_2$
$N_S(t_9) = [\{x_2, x_6, x_8, x_{10}\}, \{x_1, x_3, x_4, x_5, x_7, x_9\}]$	*marked positive* as $sup = 3$ and $conf = 0.5$
$N_S(t_{10}) = [\{x_2, x_6, x_8, x_{10}\}, \{x_2, x_6, x_8, x_{10}\}]$	not marked as $sup = 3$ but $conf = 0.18 < \lambda_2$

Second Loop:
Building action terms of length two from all possible unmarked atomic action terms.

$N_S(t_1 * t_4) = [\{x_1, x_8\}, \{x_1, x_8\}] = N_S(t_4)$	*marked but no rule*
$N_S(t_1 * t_5) = [\{\emptyset\}, \{\emptyset\}]$	*marked negative* as $card(Y_1) = 0$
$N_S(t_1 * t_6) = [\{x_9\}, \{x_9\}]$	*marked negative* as $card(Y_2 \cap Z_2) = 0$
$N_S(t_1 * t_7) = [\{x_1, x_9\}, \{x_1, x_9\}]$	*marked negative* as $card(Y_2 \cap Z_2) = 0$
$N_S(t_1 * t_8) = [\{x_1, x_9\}, \{x_8\}]$	*marked negative* as $sup = 2$ and $conf = 1$
$N_S(t_1 * t_{10}) = [\{x_8\}, \{x_8\}]$	*marked negative* as $card(Y_1 \cap Z_1) = 0$
$N_S(t_2 * t_4) = [\{\emptyset\}, \{\emptyset\}]$	*marked negative* as $card(Y_1) = 0$
$N_S(t_2 * t_5) = [\{x_5, x_6\}, \{x_5, x_6\}]$	not marked as $Y_1 = Y_2$
$N_S(t_2 * t_6) = [\{x_2, x_3, x_{10}\}, \{x_2, x_3, x_{10}\}]$	not marked as $Y_1 = Y_2$
$N_S(t_2 * t_7) = [\{x_3, x_5\}, \{x_3, x_5\}]$	*marked negative* as $card(Y_1 \cap Z_1) = 0$
$N_S(t_2 * t_8) = [\{x_3, x_5\}, \{x_2, x_6, x_{10}\}]$	*marked negative* as $card(Y_1 \cap Z_1) = 0$
$N_S(t_2 * t_{10}) = [\{x_2, x_6, x_{10}\}, \{x_2, x_6, x_{10}\}]$	*marked negative* as $card(Y_2 \cap Z_2) = 0$
$N_S(t_4 * t_7) = [\{x_1\}, \{x_1\}]$	*marked negative* as $card(Y_2 \cap Z_2) = 0$
$N_S(t_4 * t_8) = [\{x_1\}, \{x_8\}]$	*marked negative* as $sup = 1$
$N_S(t_4 * t_{10}) = [\{x_8\}, \{x_8\}]$	*marked negative* as $card(Y_1) = 0$
$N_S(t_5 * t_7) = [\{x_4, x_5, x_7\}, \{x_5, x_7\}]$	*marked negative* as $card(Y_1 \cap Z_1) = 0$
$N_S(t_5 * t_8) = [\{x_4, x_5, x_7\}, \{x_6\}]$	*marked negative* as $card(Y_2 \cap Z_2) = 0$
$N_S(t_5 * t_{10}) = [\{x_6\}, \{x_6\}]$	*marked negative* as $card(Y_2 \cap Z_2) = 0$
$N_S(t_6 * t_7) = [\{x_3, x_9\}, \{x_3, x_9\}]$	*marked negative* as $sup = 1$
$N_S(t_6 * t_8) = [\{x_3, x_9\}, \{x_2, x_{10}\}]$	*marked negative* as $card(Y_2 \cap Z_2) = 0$
$N_S(t_6 * t_{10}) = [\{x_2, x_{10}\}, \{x_2, x_{10}\}]$	*marked negative* as $card(Y_2 \cap Z_2) = 0$
$N_S(t_7 * t_{10}) = [\{\emptyset\}, \{\emptyset\}]$	*marked negative* as $card(Y_1) = 0$
$N_S(t_8 * t_{10}) = [\{\emptyset\}, \{x_2, x_6, x_8, x_{10}\}]$	*marked negative* as $card(Y_1) = 0$

Third Loop:
Building action terms of length three, four, etc from all possible unmarked shorter terms. It is repeated until we reach the fix point. In our example the

algorithm stops, as we cannot form any other action terms. Two rules we obtained are given below:

$r_1 = [(b, b_2 \rightarrow b_1)] \rightarrow (d, H \rightarrow A)$ with $sup(r_1) = 3$ and $conf(r_1) = 0.5$
$r_2 = [(a, a_1) * (b, b_1 \rightarrow b_2)] \rightarrow (d, H \rightarrow A)$ with $sup(r_2) = 2$ and $conf(r_2) = 1$.

3.5 Cost and Feasibility

In this book we also propose a method for measuring the level of re-classification freedom for objects in a decision system. Let us now introduce the notion of a cost and feasibility of an action rule. They are needed to understand the value [40], [41], [42] of action rules and the same to exploit this information for the benefits.

So, let us assume that S is an information system. Let $b \in B$ is flexible attribute and b_1, b_2 are its two values. By $\zeta(b_1, b_2)$ we mean any number from the open interval $(0, 1) \cup \{+\infty\}$ which describes the cost to change the value from b_1 to b_2 by the user of the information system S.

- The value of $\zeta(b_1, b_2) \approx 0$ is interpreted that the change of values from b_1 to b_2 is quite trivial
- The value of $\zeta(b_1, b_2) \approx 1$ is interpreted that the change of values from b_1 to b_2 is very difficult to be achieved
- The value of $\zeta(b_1, b_2) \approx +\infty$ is interpreted that the change is not feasible

Also, if $\zeta(b_1, b_2) < \zeta(b_3, b_4)$, then change of values from b_1 to b_2 is more feasible than the change from b_3 to b_4.

The values $\zeta(b_i, b_j)$ are given by the user of information system and they should be seen as atomic values needed to introduce the notion of the feasibility of an action rule.

Assume now that $r = [(b_1, v_1 \rightarrow w_1) * (b_2, v_2 \rightarrow w_2) * \dots * (b_m, v_m \rightarrow w_m)](x) = (d, d_1 \rightarrow d_2)(x)$ is a (r_1, r_2)-action rule [6], [43].

Definition 3.21. *By the cost of rule r denoted by $cost(r)$ we mean the value*

$$cost(r) = \sum \{\zeta(v_i, w_i) : 1 \leq i \leq n\}.$$

We say that the rule r is feasible if $cost(r) < \zeta(d_1, d_2)$, which means that $cost(r)$ has to be a finite number and the cost of the conditional part of the rule has to be lower than the cost of the decision part of the rule.

Now, if $\{d\}$ is a decision attribute we want to re-classify some objects (e.g. patients) in S from the group described as d_1 to the group described as d_2, then we should look for a rule with the smallest cost value. To be more precise, let us assume that $D_S[(d, d_1 \rightarrow d_2)]$ denotes the set of all action rules in S having the term $(d, d_1 \rightarrow d_2)$ on their decision site. Among all action rules in $D_S[(d, d_1 \rightarrow d_2)]$ we have to choose a rule with the smallest cost value. However, it can still happen that the rule we chose has the cost value not acceptable by the user of the information system S. The cost of the action rule

$$r_i = [(b_1, v_1 \rightarrow w_1) * (b_2, v_2 \rightarrow w_2) * ... * (b_m, v_m) \rightarrow w_m)](x_i) \rightarrow (d, d_1 \rightarrow d_2)(x_i)$$

might be high only because the cost value of one of its sub-terms in the conditional part of the rule is high.

Let us assume that $(b_i, v_i \rightarrow w_i)$ is that term. In such case, we may look for an action rule in $D_S[(b_i, v_i \rightarrow w_i)]$ with the smallest cost value. Assume that

$$r_j = [(b_{i1}, v_{i1} \rightarrow w_{i1}) * (b_{i2}, v_{i2} \rightarrow w_{i2}) * ... * (b_{in}, v_{in} \rightarrow w_{in})](x_j) \rightarrow (b_i, v_i \rightarrow w_i)(x_j)$$

is feasible rule. Since objects x_i, x_j are coming from the same information system S, we can compose r_i with r_j getting a new feasible rule given below:

$$(r_i, r_j) = [(b_1, v_1 \rightarrow w_1) * (b_{i1}, v_{i1} \rightarrow w_{i1}) * (b_{i2}, v_{i2} \rightarrow w_{i2}) * ...$$

$$... * (b_{in}, v_{in} \rightarrow w_{in}) * ... * (b_m, v_m \rightarrow w_m)](x) \rightarrow (d, d_1 \rightarrow d_2)(x).$$

The cost of this new action rule (r_i, r_j) is lower than the cost of r_i. However, if support of this rule is equal to 0, then it has no value for user. Otherwise, we can recursively follow this method looking for cheaper rules reclassifying objects from the group d_1 into the group d_2. Each successful step will produce a rule which is cheaper than the previous one. Obviously, this heuristic procedure has to end.

One can argue that if $D_S[(d, d_1 \rightarrow d_2)]$ contains all action rules which reclassify objects from one group d_1 into the group d_2 then any new action rule obtained as the result of the proposed recursive strategy is already in that set. We agree with this statement but practically $D_S[(d, d_1 \rightarrow d_2)]$ never contains all such rules.

Firstly, it is too expensive to generate all possible rules from an information system and secondly even if we have such rules it is still too expensive to generate all possible action rules from them. So the applicability of the above proposed heuristic strategy is highly justified.

Now, let us assume a slightly different scenario. The action rule

$$r_i = [(b_1, v_1 \rightarrow w_1) * (b_2, v_2 \rightarrow w_2) * ...$$

$$... * (b_m, v_m \rightarrow w_m)](x_i) \rightarrow (d, d_1 \rightarrow d_2)(x_i)$$

is extracted from the information system S and either it is not feasible because of the term $(b_i, v_i \rightarrow w_i)$ which has a very high $\zeta(v_i, w_i)$ assign to it or we are just not satisfied with $cost(r_i)$. In both cases we are looking for a new feasible action rule

$$r_j = [(b_{i1}, v_{i1} \rightarrow w_{i1}) * (b_{i2}, v_{i2} \rightarrow w_{i2}) * ...$$

$$... * (b_{in}, v_{in} \rightarrow w_{in})](x_j \rightarrow (b_i, v_i \rightarrow w_i)(x_j)$$

which can be applied to r_i to decrease its cost value. The current setting looks the same to the one we already had but this time we assume that r_j is extracted from another information system. For simplicity reason, we assume that the semantics and the granularity levels of all attributes common for both information systems are the same.

Let us consider the compositions of the action rule r_j with the rule r_i. The resulting new feasible action rule (r_i, r_j) has the form:

$$(r_i, r_j) = [(b_1, v_1 \to w_1) * (b_{i1}, v_{i1} \to w_{i1}) * (b_{i2}, v_{i2} \to w_{i2}) * \ldots$$

$$\ldots * (b_{in}, v_{in} \to w_{in}) * \ldots * (b_m, v_m \to w_m)](x) \to (d, d_1 \to d_2)(x)$$

where x is an object in S. Some of the attributes from $\{b_{i1}, b_{i2}, \ldots, b_{im}\}$ may not belong to S. Also, the rule $t_1 = (a, a_1 \to a_1)e\ r_j$ is supported by objects belonging to the information system from which r_j was extracted. Let us

Table 3.9

	b_1	b_2	b_3	b_4	b_5	b_6
x	v_1	v_2	v_3	v_4		
y	w_1	w_2	w_3		w_5	w_6

denote this system by $S_j = (X_j, B_j, V_j, f_j)$ and the set of objects in X_j supporting r_j by $sup(r_j)$. Assume that $sup(r_i)$ is the set of objects in S supporting rule r_i. The domain of (r_i, r_j) is the same as the domain of r_i which is equal to $sup(r_i)$. To define the notion of a similarity let us assume that $B = \{b_1, b_2, b_3, b_4\}$, $B_j = \{b_1, b_2, b_3, b_5, b_6\}$ and objects $x \in X, y \in X_j$ are defined by the Table 3.9.

The similarity $sim(x, y)$ between x and y is equal to:

$$\frac{1 + 0 + 0 + \frac{1}{2} + \frac{1}{2} + \frac{1}{2}}{6} = \frac{5}{12}.$$

In a formal, more general way:

Definition 3.22. *The similarity between two objects belonging to two different information systems is calculated as:*

$$sim(x, y) = \frac{\sum[sim(b_i(x), b_i(y)) : i \in I]}{card(I)},$$

where:

- $sim(b_i(x), b_i(y)) = 0$ *if* $b_i(x) \neq b_i(y)$
- $sim(b_i(x), b_i(y)) = 1$ *if* $b_i(x) = b_i(y)$
- $sim(b_i(x), b_i(y)) = \frac{1}{2}$ *if* $b_i(x)$ *or* $b_i(y)$ *is undefined.*

To find the confidence associated with the action rule (r_i, r_j) we find first $sim(x, X_j) = max\{sim(x, y) : y \in X_j\}$ for every $x \in X$. By the confidence of (r_i, r_j) we mean $conf(r_i, r_j) = \frac{\sum[sim(x, X_j:x \in X)]}{card(X)} \cdot conf(r_i)$, where $conf(r_i)$ is the confidence of the rule r_i.

If we allow building action rules from rules extracted at many sites of *DIS*, we increase our chance to construct action rules of very low cost value but at a price of their decreased confidence. Assuming that the results of treatment is the decision attribute in the information system S, rules discovered in S can be seen as definitions of different levels of results of treatment. Action rules constructed from them show different future possibilities for patients in decision system S.

3.6 Association Action Rules

The task of association rule mining is to find certain association relationships among a set of objects in large databases. The association relationships are described as rules. In [32] different measures of interestingness of a rule is described. In our approach two main measurements are calculated: support and confidence. Support corresponds to statistical significance while confidence describes the strength of the rule. Both of them however are not directly linked in the context. To resolve this problem we can consider a framework for value added association rules by attaching numerical values of itemsets, representing profits, importance, and benefits [24]. The task of discovering association rules was first introduced in [2]. In literature of data mining there are a lot of papers on designing scalable algorithms for mining association rules [3], [12], [15], [56] [72]. They are useful with a large amount of data in order to understand, e.g. the customer behavior in stores. But they are useful not only for the business sector. There have been researches examining inventory control or medical problems.

In [68] association rules are used to generate recommendation action. In their model for a given set of transactions and pre-selected target items, Wang et all suggest how to build a new model for promotion strategies to new customers, with the goal of maximizing the profit. In [27], [69] authors propose the method of generation interesting patterns by incorporating knowledge in the process of searching for patterns in data. They focus on providing methods which generate unexpected patterns with respect to intuition.

Next they show a new approach for generating association-type action rules. The notion of frequent action sets and Apriori-like strategy generating them is proposed. Finally, the notion of a representative action rules is given and a method to construct them directly from frequent action sets is presented.

We propose a slightly different approach to achieve the following objectives:

1. Extract action rules directly from a decision system without using pre-existing classification rules.
2. Extract action rules that have minimal attribute involvement.

To meet these two goals, we introduce the notion of action frequent action sets, and present methods of building action rules from them.

Now, let us assume that $S = (X, A, V)$ is an information system with two thresholds: λ_1, λ_2 describing minimum support and minimum confidence assigned to action rules, respectively. The algorithm for constructing frequent action sets is similar to Agrawal's algorithm [3].

3.6.1 Frequent Action Sets

Let t_a is an atomic action set, where $N_S(t_a) = [Y_1, Y_2]$ and $a \in A$. We say that t_a is called frequent if $card(Y_1) \geq \lambda_1$ and $card(Y_2) \geq \lambda_2$.

The operation of generating $(k + 1)$- element candidate action sets from frequent k-element action sets is performed in two steps:

1. **Merging Step:** Merge pairs (t_1, t_2) of frequent k-element action sets into $(k + 1)$-element candidate action set if all elements in t_1 and t_2 are the same except the last element.
2. **Pruning Step:** Delete each $(k+1)$-element candidate action set t if either it is not an action set or some k-element subset of f is not a frequent k-element action set.

Now, if t is a $(k + 1)$-element candidate action set, $N_S(t) = [Y_1, Y_2]$, $card(Y_1) \geq \lambda_1$ and $card(Y_2) \geq \lambda_2$, then t is a frequent $(k + 1)$-element action set. We say that t is a frequent action set in S if t is a frequent k-element action set in S, for some k.

Assume now that the expression $[t - t_1]$ denotes the action set containing all atomic action sets listed in t but not listed in t_1. The set $AARS(\lambda_1, \lambda_2)$ of association action rules in S is constructed in the following way:

Let t be a frequent action set in S and t_1 is its subset. Any action rule $r = [(t - t_1) \rightarrow t_1]$ is an association action rule in $AARS(\lambda_1, \lambda_2)$ if $conf(r) \geq \lambda_2$.

Example 3.23. Let us take information system the same as in Table 2.4. We take $\lambda_1 = 2, \lambda_2 = 0.4$. The following frequent action sets can be constructed:

(a, a_1)	support 3
(a, a_2)	support 5
(a, a_3)	support 2
(b, b_1)	support 6
(b, b_2)	support 4
$(b, b_1 \rightarrow b_2)$	support 6
$(b, b_2 \rightarrow b_1)$	support 4
(c, c_1)	support 2
(c, c_2)	support 4
(c, c_3)	support 4
(d, H)	support 5
(d, A)	support 5

Building pairs from all possible atomic terms:

$(a, a_1) * (b, b_1)$	support 2
$(a, a_1) * (b, b_2)$	support 1 (*not frequent*)
$(a, a_1) * (b, b_1 \rightarrow b_2)$	support 2
$(a, a_1) * (b, b_2 \rightarrow b_1)$	support 1 (*not frequent*)
$(a, a_1) * (c, c_1)$	support 2
$(a, a_1) * (c, c_2)$	support 0 (*not frequent*)
$(a, a_1) * (c, c_3)$	support 1 (*not frequent*)
$(a, a_1) * (d, H)$	support 2
$(a, a_1) * (d, A)$	support 1 (*not frequent*)
$(a, a_2) * (b, b_1)$	support 2
$(a, a_2) * (b, b_2)$	support 3
$(a, a_2) * (b, b_1 \rightarrow b_2)$	support 2
$(a, a_2) * (b, b_2 \rightarrow b_1)$	h support 3
$(a, a_2) * (c, c_1)$	support 0 (*not frequent*)
$(a, a_2) * (c, c_2)$	support 2
$(a, a_2) * (c, c_3)$	support 3
$(a, a_2) * (d, H)$	support 3
$(a, a_2) * (d, A)$	support 2
...	...
$(b, b_1) * (c, c_1)$	support 1 (*not frequent*)
$(b, b_1) * (c, c_2)$	support 3
$(b, b_1) * (c, c_3)$	support 2
...	..
$(a, a_2) * (b, b_2 \rightarrow b_1) * (c, c_2) * (d, H \rightarrow A)$	support 2

Association action rules can be constructed from all frequent action sets. For instance, we can generate association action rule $[(a, a_2) * (b, b_2 \rightarrow b_1) * (c, c_2)] \rightarrow (d, H \rightarrow A)$ from the last frequent action set listed above. We can also construct simple association action rule, calculate the cost of association action rule, and give a strategy to construct simple association action rules of lowest cost.

3.7 Representative Association Action Rules

The concept of representative association rules was introduced by Kryszkiewicz [23]. They form a small subset of association rules from which the remaining association rules can be generated. Similar approach for action rules was also presented in [48], [55].

Definition 3.24. *By a cover of association rule* $r : (t_1 \rightarrow t)$ *we mean* $cov(r) = cov(t_1 \rightarrow t) = \{t_1 * t_2 \rightarrow t_3 : t_2, t_3$ *are not overlapping subterms of* $t\}$.

Let us assume that $r : [(a, a_1 \to a_2) \to (b, b_1 \to b_2) * (c, c_1 \to c_2) * (d, d_1 \to d_2)]$ is an association rule. Then, $((a, a_1 \to a_2) * (b, b_1 \to b_2) \to (c, c_1 \to c_2)) \in cov(r)$.

Property 1. If $r \in AARS(\lambda_1, \lambda_2)$, then each rule $r_k \in cov(r)$ also belongs to $AARS(\lambda_1, \lambda_2)$.

Proof: From the definition of $AARS(\lambda_1, \lambda_2)$ we have: $sup(r) \geq \lambda_1$ and $conf(r) \geq \lambda_2$. Let $r_k = (t_1 * t_2) \to t_4$, $r = (t_1 \to t_2) * t_3 * t_4$, and $N_S(t_i) = [Y_i, Z_i]$ for $i \in \{1, 2, 3, 4\}$. Since $\frac{card[Y_1 \cap Y_2 \cap Y_3 \cap Y_4]}{card[Y_1]} \geq \lambda_1$ then $\frac{card[Y_1 \cap Y_2 \cap Y_4]}{card[Y_1 \cap Y_2]} \geq \lambda_1$. It comes from the fact, that $card[Y_1 \cap Y_2 \cap Y_4] \geq card[Y_1 \cap Y_2 \cap Y_3 \cap Y_4]$ and $card[Y_1] \geq card[Y_1 \cap Y_2]$.

In a similar way we show that $\frac{card[Z_1 \cap Z_2 \cap Z_4]}{card[Z_1 \cap Z_2]} \geq \lambda_1$.

The same, $sup(r_k) \geq \lambda_1$.

Now assume, that $conf(r) = \frac{card[Y_1 \cap Y_2 \cap Y_3 \cap Y_4]}{card[Y_1]} \cdot \frac{card[Z_1 \cap Z_2 \cap Z_3 \cap Z_4]}{card[Z_1]} \geq \lambda_2$.

Clearly, $\frac{card[Y_1 \cap Y_2 \cap Y_4]}{card[Y_1 \cap Y_2]} \cdot \frac{card[Z_1 \cap Z_2 \cap Z_4]}{card[Z_1 \cap Z_2]} \geq \lambda_2$.

The same $conf(r_k) \geq \lambda_2$.

Property 2. Representative association rules $RAARS(\lambda_1, \lambda_2)$ form a least set of representative association action rules that covers all association action rules $AARS(\lambda_1, \lambda_2)$

Proof: Assume that $r \in RAARS(\lambda_1, \lambda_2)$ and there exists $r_k \in (t_1 \to t) \in AARS(\lambda_1, \lambda_2)$ such that $r_k \neq r$ and $r \in cov(r_k)$. Since $r \in cov(r_1)$ then r is not in $RRARS(\lambda_1, \lambda_2)$.

Property 3. All association rules $AARS(\lambda_1, \lambda_2)$ can be derived from representative association action rules $RAASS(\lambda_1, \lambda_2)$ by means of cover operator.

Proof: Assume that $r : (t \to s) \in AARS(\lambda_1, \lambda_2)$ and $t = t_1 * t_2 * \ldots * t_n$, where $\{t_i\}_{i \in \{1, 2, \ldots, n\}}$ are atomic action sets. It means, that $sup(r) \geq \lambda_1$ and $conf(r) \geq \lambda_2$. Let $r_i(t) = ((t - t_i) \to s * t_i)$ for any atomic action set t_i in t. Clearly, $sup(r_i(t)) = sup(r)$ and $conf(r_i(t)) \leq conf(r)$.

Now we show how to construct representative association action rule from which r can be generated. It consists of two main steps: First we:

1. find t_i in t such that $conf(r_i(t)) \geq \lambda_2$
2. if succeeded then $t := (t - t_i)$, $s := s * t_i$ and we go back to (1). Otherwise procedure stops

In next step we extend the decision part of the rule generated in previous step. Assuming that $t \to s$ is such rule, and $T = \{t_1, t_2, \ldots, t_m\}$ is a set of all atomic action terms not listed in s, we:

1. find $t_i \in T$ such that $sup(t \to s * t_i) \geq \lambda_1$
2. if succeeded then $T := T - t_i$, $s := s * t_i$ and we go back to (1). Otherwise procedure stops

The resulting association action rule is a representative rule from which the initial rule r can be generated.

3.8 Simple Association Action Rules

In this section we introduce the notion of a simple association action rule and give an idea how to construct simple association action rules of lowest cost. Let $(a, a_1 \rightarrow a_2)$ is an atomic action set. We assume that the cost of changing attribute a from a_1 to a_2 is denoted by $cost_S((a, a_1 \rightarrow a_2))$ [64], [65]. For simplicity reason, the subscript S will be omitted.

Let $t_1 = (a, a_1 \rightarrow a_2)$, $t_2 = (b, b_1 \rightarrow b_2)$ be two atomic action sets. We say that t_1, t_2 are positively correlated if change t_1 supports change t_2. Saying another words, change t_1 implies change t_2.

Assume that action set t is constructed from atomic action sets $T = \{t_1, t_2, ..., t_m\}$. We introduce a binary relation $\cong$ on T defined as: $t_i \cong t_j$ iff t_i and t_j are positively correlated. Relation $\cong$ is an equivalence relation and it partitions T into M equivalence classes $(T = T_1 \cup T_2 \cup ... \cup T_M)$.

In each equivalence class T_i, an atomic action set $a(T_i)$ of the lowest cost is identified. The cost of t is defined as:

$$cost(t) = \{\sum cost(a(T_i)) : 1 \leq i \leq m\}.$$

Now, assume that $r = (t_1 \rightarrow t)$ is an association action rule. We say that r is simple if $cost(t_1 * t) = cost(t_1)$. The cost of r is defined as $cost(t_1)$. We assume that user gives three threshold values, λ_1 - minimum support, λ_2 - minimum confidence, λ_3 - maximum cost. Let t be a frequent action set in S and t_1 is its subset.

Any association action rule $r \in AARS_S(\lambda_1, \lambda_2)$ is called *association action rule of acceptable cost* if $cost(r) \leq \lambda_3$.

Similarly, frequent action set t is called a *frequent action set of acceptable cost* if $cost(t) \leq \lambda_3$.

In order to construct simple association action rules of a lowest cost, we built frequent action sets of acceptable cost following the strategy presented in Section 3.6 enhanced by additional constraint which requires to verify the cost of frequent action sets being produced. Any frequent action set which cost is higher than λ_3, is removed. Assuming that its confidence is not greater than λ_2, if t is a frequent action set of acceptable cost and $\{a(T_i) : i \leq m\}$ is a collection of atomic action sets constructed by following the strategy presented in this section, then $\prod\{a(T_i) : i \leq m\} \rightarrow (t - \{a(T_i) : i \leq m\})$ is a simple association action rule of acceptable cost.

3.9 Action Reducts

In this section we present a method that provides a simple set of attribute values which need to be modified in order to shift a class of objects from one class to another. This method is based on action reducts introduced in [19]. The recommendations made by action reduct are quite simple, i.e. change a couple of values to have a better prognoses.

Suppose that the patients from rehabilitation ward can be classified into several groups according to their medical treatment levels, such as high, neutral, or low. One thing the hospital can do to improve the cure is finding a way to make the patients healthier and more able so that they continue to do exercises in rehabilitation ward. The algorithm described in this section tries to solve such problem using existing data. Assume that a hospital maintains a database for patient information in a table. The table has columns describing the characteristics of each patient, such as personal information, earlier illnesses, laboratory analysis etc. We divide the patients into two groups based on the medical treatment level, which is a decision value. The first group is comprised of satisfied patients who will keep doing their exercises for an extended period time. The second group is comprised of neutral or unsatisfied patients. Our goal is to find a set of distinct values or unique patterns from the first group that do not exist in the second group. These characteristics of the satisfied and healthier patients can be used by the hospital to improve the patients conditions for the people in the second group. Some of attribute values describing the patients can be controlled or changed, which is defined as an action.

We propose the action reduct, in this section, to formulate necessary actions.

An action reduct has following properties:

1. It is the minimal set.
2. It is obtained from objects having favorable decision values.
3. It is a distinct set of values not found in the other group, the group not having favorable decision values.

There are four attributes, $\{a, b, c, d\}$. We classify the attributes into two types: condition and decision. The condition attributes are $\{a, b, c\}$, and the decision attribute is $\{d\}$. We assume that the set of condition attributes is further

Table 3.10 Incomplete Information System S_1 of type $\lambda = 0.3$

X	Attribute a	Attribute b	Attribute c	Attribute d
x_1	a_2	b_1	c_1	A
x_2	a_1	b_3	c_2	A
x_3	a_1	b_1		A
x_4	a_1	b_3	c_1	A
x_5	a_1	b_1	c_1	H
x_6	a_1	b_1	c_1	H
x_7	a_2		c_2	H
x_8	a_1	b_2	c_2	H

partitioned into stable attributes A_{St} and flexible attributes A_{Fl}. In our example $A_{St} = \{c\}$, $A_{Fl} = \{a, b\}$. The values in decision attribute d are divided into two sets:

- $d_\alpha = \{v_i \in V_d : v_i \text{ is a desired decision value}\}$
- $d_\beta = V_d \setminus \{d_\alpha\}$

For simplicity we use an example having only one element d_2 in d_α and d_1 in d_β. However, the algorithm presented in this section carries over to the general case.

The objects are partitioned into two groups based on the decision values.

- $X_\alpha = \{x_i \in X : d(x_i) \in d_\alpha\}$ (objects with decision value in d_α)
- $X_\beta = \{x_j \in X : d(x_j) \in d_\beta\}$ (objects with decision value in d_β)

In Table 3.10, $X_\alpha = \{x_1, x_2, x_3, x_4\}$, $X_\beta = \{x_5, x_6, x_7, x_8\}$. We want to provide the user a list of attribute values that can be used to make changes on some of the objects to steer the unfavorable decision value to a more favorable value. We use the reduct [30], [58] to create the list. By a reduct relative to an object x we mean a minimal set of attribute values distinguishing x from all other objects in the information system [19]. For example, $\{a_2, c_2\}$ is a reduct relative to x_7 since $\{a_2, c_2\}$ can differentiate x_7 from other objects in S.

To construct $\alpha - reduct$ we partitioned the objects in S into two groups according to their decision values. Objects in X_α have d_2 that is the favorable decision value. Our goal is to identify which sets of condition attribute values describing objects in X_α make them different from the objects in X_β.

Table 3.11 $\alpha - reduct$ for information system S

α-reduct	weight w	frequency f	Hit ratio h
$\{a_2, b_1\}$	0.29	1	1
$\{a_2, c_1\}$	0.14	1	0.5
$\{b_3\}$	0.57	2	1

The notion of partial uniqueness and significant number will be introduced to measure the usability of $\alpha-reducts$. In Table 3.11 there are $\alpha-reducts$ for S. Those are the smallest sets of condition attribute values that are different from the condition attribute values representing objects in X_β. We obtained the first two $\alpha-reducts$ relative to x_1. These are the prime implicants [11] of the differences between x_1 and $\{x_5, x_6, x_7, x_8\}$. Subsequent $\alpha - reducts$ are extracted using objects x_2, x_3, x_4 for S.

We need a method to measure the usability of the α-reducts [19]. It is possible that more than one object can have the same $\alpha - reduct$. The frequency of an $\alpha - reduct$ in X_α is defined as f. The $\alpha - reduct$ may not be used to make changes for some of the attribute values of $x \in X_\beta$. That is because some objects in X_β do not differ in terms of their attribute values. We also are not able to modify stable attribute values. The hit ratio, represented as h, is the ratio between the number of applicable objects and the total number of objects in X_β. An applicable object is the object that the attribute values are different from those in $\alpha - reduct$, and they are not stable values. We define the function to measure the usability.

The weight of an $\alpha - reduct$ k is $w_k = \frac{f_k * h_k}{\sum (f * h)}$, where f_k and h_k are the frequency and hit ratio for k, and $\sum (f * h)$ is the sum of the weights of all $\alpha - reducts$. It provides a way to prioritize the $\alpha - reduct$ using a normalized value.

Table 3.12 X_α Set of objects with decision value d_A

X	Attribute a	Attribute b	Attribute c	Attribute d
x_1	a_2	b_1	c_1	A
x_2	a_1	b_3	c_2	A
x_3	a_1	b_1		A
x_4	a_1	b_3	c_1	A

Table 3.13 X_β Set of objects with decision value d_H

X	Attribute a	Attribute b	Attribute c	Attribute d
x_5	a_1	b_1	c_1	H
x_6	a_1	b_1	c_1	H
x_7	a_2		c_2	H
x_8	a_1	b_2	c_2	H

Let us back to the example. Using the partitions in Table 3.12 and Table 3.13, we extract the distinct attribute values of $x \in X_\alpha$ against $x \in X_\beta$. The following matrix shows the discernable attribute values for objects $\{x_1, x_2, x_3, x_4\}$ against objects $\{x_5, x_6, x_7, x_8\}$.

Table 3.14 Matrix for $\{x_1, x_2, x_3, x_4\}$ against $\{x_5, x_6, x_7, x_8\}$

	x_1	x_2	x_3	x_4
x_5	a_2	$b_3 + c_2$	$\emptyset$	b_3
x_6	a_2	$b_3 + c_2$	$\emptyset$	b_3
x_7	$b_1 + c_1$	$a_1 + b_3$	$a_1 + b_1$	$a_1 + b_3 + c_1$
x_8	$a_2 + b_1 + c_1$	b_3	b_1	$b_3 + c_1$

For example, a_2 in x_1 is different from x_5. We need a_2 to discern x_1 from x_5. Either b_1 or ("+") c_1 can be used to distinguish x_1 from x_7. In order to find the minimal set of values that distinguishes x_1 from all objects in $X_\beta = \{x_5, x_6, x_7, x_8\}$ we multiply all discernable values: $(a_2) * (a_2) * (b_1 + c_1) * (a_2 + b_1 + c_1)$. That is, (a_2) and (b_1 or c_1) and (a_2 or b_1 or c_1) should be different to make x_1 distinct from all other objects [30], [58]. The $\alpha - reduct$ relative to x_1, $r_\alpha(x_1)$, using the conversion and the absorption law is:

$r_\alpha(x_1) = (a_2) * (a_2) * (b_1 + c_1) * (a_2 + b_1 + c_1) = (a_2) * (b_1 + c_1) = (a_2 * b_1) + (a_2 * c_1)$. A missing attribute value of an object in X_α does not qualify to discern the object from the objects in X_β because it is undefined. A missing value in X_β is regarded as a different value if a value is present in $x \in X_\alpha$. When a discernible value does not exist, we do not include it in the calculation of the prime implicant. We acquired the following $\alpha - reduct$ relative to objects in X_β.

$r_\alpha(x_1) = (a_2 * b_1) + (a_2 * c_1)$
$r_\alpha(x_2) = (b_3)$
$r_\alpha(x_3) = $NULL
$r_\alpha(x_4) = (b_3)$

After calculating the $\alpha - reducts$, we have to measure their usability. The frequency of $\alpha - reduct$ $\{a_2, b_1\}$ is 1 because it appears in X_α once (Table 3.11). The hit ratio is $\frac{4}{4} = 1$, which means that we can use the reduct for all objects in X_β. The weight is $w = \frac{f*h}{\sum(f*h)} = \frac{2}{7} = 029$.

The values in the stable attribute c cannot be modified. The hit ratio for $\alpha - reduct$ $\{a_2, c_1\}$ is $\frac{2}{4} = 0.5$ because we cannot re-classify objects x_7 and x_8 from (d, H) to (d, A). The stable value c_2 in the two objects cannot be changed. Using the $\alpha - reduct$ $\{b_3\}$ that has the highest weight, we can make a recommendation for changing the value of attribute b in X_β to b_3 in order to induce the decision value in X_β to (d, A).

3.9.1 Experiments and Testing

Presented method was testing on children flat foot database, obtained from Bialystok Childrens' Hospital. Foot problems are reported by many children and adults. The plano-valgus is very often foot dysfunction seen, especially in children. Prediction the risk of plano-valgus provides valuable information for prevention this dysfunction and the consequences in future life. The main purpose of our experiments was a proposition of building a new system for prediction and prevention plano-valgus in children. We examined plantar arch height in 60 children with flatfoot aged from 7 to 15 years. The evaluation criteria included low arch height, correct knee and heel position, and correct body symmetry. To measure arch height, the optoelectronic system was used. The system to prediction a risk of plano-valgus in children is based on such factors as: age, Cole'index, gender, place of living, children's physical activity, joint hypermobility, Wejsfolg index, and others. Many children with plano-valgus have flexible arches during loading. When the foot is not bearing weight, it has an arch. As soon as weight is placed on the foot, the arch collapses. Children with flexible plano-valgus may have a high arched foot when non-weightbearing but pronate excessively during gait [17], [31]. This means the foot stays flat or pronated at a time when it should be supinating and becoming more rigid. This results in instability which may result in bunions, hammertoes, heel pain, arch pain, metatarsalgia and tendonitis [25], laxity, hyperextension of knees and elbows, hyperextension of wrist and fingers, and excess ankle dorsiflexion. Some of children with plano-valgus may feel foot pain, ankle pain, or lower leg pain are present, especially in children. Painful in children may be caused by a condition called tarsal coalition. Tarsal coalition is a condition where two or more of the bones in the foot fuse together, limiting motion and often leading to a flat foot [17]. Ethiopathogenesis of flat foot has not been fully explained yet, but the causes are usually connected to: age, gender, place of residence, muscular stress and occurrence of weariness, physical activity, and hypermobility. The flat foot results in changes in muscle work and the muscular strength of calves and feet may be weakened, which may affect the development of the arch of the foot during growth.It is evident that there is need for early detection and timely prevention of plano-valgus dysfunction. We hypothesized that and action rule discovery methods help us understand the relationships among the risk factors and measurements in order to better understand the cause of plano-valgus and gain new knowledge for predicting plano-valgus risk.

Table 3.15 Dataset used for the initial experiments

Attributes names	Values of attribute	Type of attributes
patient's age	{child(5-10), teenager(11-15), adult (16-18)}	Stable
prescription	{orthotic insoles, exercises}	Flexible
flat foot	{Yes, No}	Flexible
arch height	{normal, reduced}	Flexible
treatment	{exercises, insoles, operation}	Decision

The algorithm was implemented in Python 2.6, and tested using a data set from [28]. The data set contains information about methods of patients with flat foot treatment. Table 3.15 shows the names, descriptions, and the partitions of attributes.

The decision attribute, has three classes: only exercises without orthotic insoles, orthotic insoles and operation. The first class we set (*exercises, no insoles*) as the favorable decision value. The data set has four condition attributes. It was assumed that the age of the patient is a stable attribute, and prescription, flat foot and arch height are flexible ones. All attributes are

Table 3.16 α-reduct for d_α=exercises

$\alpha - reduct$	Weight w	Frequency f	Hit ratio h
(age,young), (flat foot, no), (arch height, normal)	0.12	2	0.45
(prescription,exercises), (flat foot, no), (arch height, normal)	0.78	3	1
(age,child), (flat foot, no), (arch height,normal)	0.10	2	0.45

categorical in the dataset. Table 2.16 shows action reducts generated during experiments.

The first $\alpha - reduct$ can be interpreted as: for objects with attribute (age,young), change the values in attributes flat foot, arch height to the suggested values (no, normal) in order to change the decision to "exercises".

The second $\alpha - reduct$ can be interpreted as: change the values in attributes prescription, flat foot and arch height to the suggested values (exercises, no, normal) in order to change the decision to "exercises".

The third $\alpha - reduct$ can be interpreted as: for objects with attribute (age,child) change the values in attributes prescription,flat foot and arch height to the suggested values (exercises, no, normal) in order to change the decision to "exercises".

Because there is no stable attribute value in the second $\alpha - reduct$ and the same pattern is not in X_β, its hit ratio is equal 1.

Interesting rules about the relationship speed of partial recovery against symptoms, rehabilitation therapy parameters, and other factors were revealed during the experiments. Also, during the experiments we observed that the new temporal features for continuous data gave more interesting rules especially for extensively improved recovery cases, which confirmed our assumption that carefully designed temporal features may contribute to a better representation of the characteristics in a future.

3.10 Meta-action

As it was stated earlier, an action rule is a rule extracted from an information system that describes a possible transition of objects from one state to another with respect to a distinguished attribute called a decision attribute [52]. Values of flexible attributes can be changed.

Meta-actions are defined as actions which trigger changes of flexible attributes either directly or indirectly because of correlations among certain attributes in the system [10]. Links between meta-actions and changes they trigger within the values of flexible attributes can be defined by an ontology [44], citewang or by a mapping linking meta-actions with changes of attribute values used to describe objects in the decision system. In medical domain, taking a drug is a classical example of a meta-action. For instance, a drug can eliminate main symptoms of disease, but at the same time it can cause side effects for a patient. Therefore doctors have to order certain lab tests to check patients response to that particular drug.

By meta-actions associated with system S we mean higher level concepts representing actions introduced in [41], [70]. Meta-actions, when executed, are expected to trigger changes in values of some flexible attributes in S as described by influence matrix [70]. To give an example, let us assume that classification attributes in S describe evaluation of rehabilitation exercises and the decision attribute represents their overall score. Examples of classification attributes are: *Do exercises effectively, Stimulate patient during exercises, Provide sufficient feedback about conditions of patients.* Examples of meta-actions associated with S will be: *Change the the kind of exercises, Change the orthotic insoles.* The influence matrix [70] is used to describe the relationship between meta-actions and the expected changes within classification attributes. It should be mentioned here that expert knowledge concerning meta-actions involves only classification attributes. Now, if some of these attributes are correlated with the decision attribute, then any change in their values will cascade to the decision attribute through this correlation. The goal of an action rule discovery is to identify all correlations between classification attributes and the decision attribute.

Let $S = (X, A, V)$ is an information system, where $V = \bigcup \{V_a : a \in A\}$. First, we modify the notion of an atomic action set given in Section 2.4 so it may include numerical attributes.

Definition 3.25. *By an atomic action set we mean any of the three expressions:*

1. *$a, a_1 \rightarrow a_2$, where a is a symbolic attribute and $a_1, a_2 \in V_a$,*
2. *$(a, [a_1, a_2] \uparrow [a_3, a_4])$, where a is a numerical attribute, $a_1 \leq a_2 < a_3 \leq a_4$, and $\forall (i \in [1, 4] \rightarrow (a_i \in V_a)$*
3. *$(a, [a_1, a_2] \downarrow [a_3, a_4])$ where a is a numerical attribute, $a_3 \leq a_4 < a_1 \leq a_2$, and $\forall (i \in [1, 4]) \rightarrow (a_i \in V_a)$*

If attribute a is symbolic and $a_1 = a_2$, then a is called stable on a_1.

The term (a, a_1) should be interpreted as *the value of attribute a stays unchanged,* while the term $(a, a_1 \rightarrow a_2)$ should be interpreted as *the value of attribute a is changed from a_1 to a_2.*

The term $(a, [a_1, a_2] \uparrow [a_3, a_4])$ should be interpreted as the value of attribute a from to the interval $[a_1, a_2]$ is increased and it belongs to the interval $[a_3, a_4]$

The term $(a, [a_1, a_2] \downarrow [a_3, a_4])$ should be interpreted as the value of attribute a from to the interval ($[a_1, a_2]$ is decreased and it belongs to the interval $[a_3, a_4]$

We will use, as previously, a simplified version of the atomic action set, e.g. $(a, \uparrow [a_3, a_4]), (a, [a_1, a_2] \uparrow), (a, \downarrow [a_3, a_4]), and (a, [a_1, a_2] \downarrow)$.

Any collection of atomic action sets is called a *candidate action set*. If a candidate action set does not contain two atomic action sets referring to the same attribute, then it is called an *action set*.

$\{(b, b_2), (b, [b_1, b_2] \uparrow [b_3, b_4])\}$ is an example of a candidate action set which is not an action set.

By the domain of *action set* t, denoted by $Dom(t)$, we mean the set of all attribute names listed in t. For instance if $\{(a, a_2), (b, b_1 \to b_2)\}$ is the action set, then its domain is equal to $\{a, b\}$.

By an *action term* we mean a conjunction of atomic action sets forming an action set. There is some similarity between atomic action sets and atomic expressions introduced in [54], [70].

Now, assume that M_1 is a meta-action triggering the action set

$$\{(a, a_2), (b, b_1 \to b_2)\}$$

and M_2 is a meta-action triggering the atomic actions in

$$\{(a, a_2), (c, c_1 \to c_2)\}.$$

It means that M_1 and M_2 involve attributes a, b, c with attribute a remaining stable in both cases.

Consider a set of meta-actions $\{M_1, M_2, ..., M_n\}$ associated with a decision system $S = (X, A, V \cup \{d\})$. Each *meta-action* M_i can invoke changes of some attribute values for objects in S. We assume here that $A \setminus \{d\} = \{A_1, A_2, ..., A_m\}$. The influence of a meta-action M_i on attribute A_j in S is represented by an atomic action set $E_{i,j}$. The influence of meta-actions $\{M_1, M_2, ..., M_n\}$ on the classification attributes in S is described by the influence matrix $\{E_{i,j} : 1 \leq i \leq n \wedge 1 \leq j \leq m\}$.

Definition 3.26. *Standard interpretation* N_S *of action terms in system* $S = (X, A, V)$ *is defined as follow:*

1. *If* $(a, a_1 \to a_2)$ *is an atomic term, then* $N_S((a, a_1 \to a_2)) = [\{x \in X : a(x) = a_1\}, \{x \in X : a(x) = a_2\}]$
2. *If* $(a, [a_1, a_2] \downarrow [a_3, a_4])$ *is an atomic term, then* $N_S((a, [a_1, a_2] \downarrow [a_3, a_4])) = [\{x \in X : a_1 \leq a(x) \leq a_2\}, \{x \in X : a_3 \leq a(x) \leq a_4\}]$
3. *If* $(a, [a_1, a_2] \uparrow [a_3, a_4])$ *is an atomic term, then* $N_S((a, [a_1, a_2] \uparrow [a_3, a_4])) = [\{x \in X : a_1 \leq a(x) \leq a_2\}, \{x \in X : a_3 \leq a(x) \leq a_4\}]$
4. *If* $(a, \uparrow [a_3, a_4])$ *is an atomic term, then* $N_S((a, \uparrow [a_3, a_4])) = [\{x \in X : a(x) < a_3\}, \{x \in X : a_3 \leq a(x) \leq a_4\}]$
5. *If* $(a, [a_1, a_2] \uparrow)$ *is an atomic term, then* $N_S((a, [a_1, a_2] \uparrow)) = [\{x \in X : a_1 \leq a(x) \leq a_2\}, \{x \in X : a_2 < a(x)\}]$

6. If $(a, \downarrow [a_3, a_4])$ is an atomic term, then $N_S((a, \downarrow [a_3, a_4])) = [\{x \in X : a(x) < a_3\}, \{x \in X : a_3 \leq a(x) \leq a_4\}]$
7. If $(a, [a_1, a_2] \downarrow)$ is an atomic term, then $N_S((a, [a_1, a_2] \downarrow)) = [\{x \in X : a_1 \leq a(x) \leq a_2\}, \{x \in X : a_2 < a(x)\}]$
8. If $t_1 = t_2 \cdot t$ is an action term, t_2 is an atomic action set, $N_S(t_2) = [Z_1, Z_2]$ and $N_S(t) = [Y_1, Y_2]$ then $N_S(t_1) = [Z_1 \cap Y_1, Z_2 \cap Y_2]$

The support and confidence for action rules are calculated as it was mentioned earlier (Section 2.3).

Assume that $S = (X, A, V \cup \{d\})$ is a decision system, $A \setminus \{d\} = \{A_1, A_2, ..., A_m\}$, $\{M_1, M_2, ..., M_n\}$ are meta-actions associated with S, $\{E_{i,j} : 1 \leq i \leq n, 1 \leq j \leq m\}$ is the influence matrix, and $r = [(A_{[i,1]}, a_{[i,1]} \rightarrow a_{[j,1]}) \cdot (A_{[i,2]}, a_{[i,2]} \rightarrow a_{[j,2]}) \cdot ... \cdot (A_{[i,k]}, a_{[i,k]} \rightarrow a_{[j,k]})] \rightarrow (d, d_i \rightarrow d_j)$ is a candidate action rule extracted from S. Also, we assume here that $A_{[i,j]}(M_i) = E_{i,j}$. Value $E_{i,j}$ is either an atomic action set or NULL. By meta-actions based decision system, we mean a triple consisting with S, meta-actions associated with S, and the influence matrix linking them.

We say that r is valid in S with respect to meta-action M_i, if the following condition holds:

- if $(\exists p \leq k) [A_{[i,p]}(M_i)$ is defined], then $(\forall p \leq k)$ [if $A_{[i,p]}(M_i)$ is defined, then $(A_{[i,p]}, a_{[i,p]}) \rightarrow a_{[j,p]} = (A_{[i,p]}, E_{i,p})$]

We say that r is valid in S with respect to meta-actions $\{M_1, M_2, ..., M_n\}$, if there is $\{i, 1 \leq i \leq n\}$, such that r is valid in S with respect to meta-action M_i.

Table 3.17 Information system S

X	Attribute a	Attribute b	Attribute c	Attribute d
x_1	a_1	b_1	c_1	H
x_2	a_2	b_1	c_2	H
x_3	a_2	b_2	c_2	H
x_4	a_2	b_1	c_1	H
x_5	a_2	b_3	c_2	H
x_6	a_1	b_1	c_2	A
x_7	a_1	b_2	c_2	H
x_8	a_1	b_2	c_1	L

To give an example, assume that S is a decision system represented by Table 3.17.

Example 3.27. Attributes $\{a, c, d\}$ are flexible and $\{b\}$ is stable. Expressions $(a, a_2), (b, b_2), (c, c_1 \rightarrow c_2), (d, d_1 \rightarrow d_2)$ are examples of atomic action sets.

Expression $r = [[(a, a_2) * (c, c1 \rightarrow c2)] \rightarrow (d, H \rightarrow A)]$ is an example of an action rule. The rule says that if value a_2 remains unchanged and value c will change from c_1 to c_2, then it is expected that the value d will change from H to A. The domain $Dom(r)$ of action rule r is equal to $\{a, c, d\}$.

We can find a number of action rules associated with S. Let us take $r = [((b, b_1) * (c, c_1 \to c_2))] \to (d, H \to A)]$ as an example of action rule. Then,

$N_S((b, b_1)) = [\{x_1, x_2, x_4, x_6\}, \{x_1, x_2, x_4, x_6\}]$,
$N_S((c, c_1 \to c_2)) = [\{x_1, x_4, x_8\}, \{x_2, x_3, x_5, x_6, x_7\}]$,
$N_S((d, H \to A)) = [\{x_1, x_2, x_3, x_4, x_5, x_7\}, \{x_6\}]$,
$N_S((b, b_1) * (c, c_1 \to c_2)) = [\{x_1, x_4\}, \{x_2, x_6\}]$.

Clearly, $sup(r) = 1$ and $conf(r) = \frac{1}{2}$.

Table 3.18 Influence matrix for S

	Attribute a	Attribute b	Attribute c
M_1		b_1	$c_2 \to c_1$
M_2	$a_2 \to a_1$	b_2	
M_3	$a_1 \to a_2$		$c_2 \to c_1$
M_4		b_1	$c_1 \to c_2$
M_5			$c_1 \to c_2$
M_6	$a_1 \to a_2$		$c_1 \to c_2$

Let $\{M_1, M_2, M_3, M_4, M_5, M_6\}$ be the set of meta-actions assigned to S with an influence matrix shown in Table 3.18. Empty slots in this table are interpreted as NULL values. In this scenario, two candidate action rules have been constructed:

$r_1 = [[(b, b_1) * (c, c_1 \to c_2)] \to (d, H \to A)$
$r_2 = [(a, a_2 \to a_1) \to (d, H \to A)]$.

The rule r_1 is valid in S with respect to M_4 and M_5. Also, rule r_2 is valid in S with respect to M_1, M_4, M_5 because there is no overlap between the domain of action rule r_2 and the set of attributes influenced by any of these meta-actions. However, we can not say that r_2 is valid in S with respect to M_2 since b_2 is not listed in the classification part of r_2.

Assume that $S = (X, A, V \cup \{d\})$ is a decision system with meta-actions $\{M_1, M_2, ..., M_n\}$ associated with S. Any candidate action rule extracted from S which is valid in a meta-actions based decision system is called action rule. So, the process of action rules discovery is simplified to checking the validity of candidate action rules.

3.10.1 Discovering Action Paths

Definition 3.28. *By an action path we mean a sequence* $[t_1 \to t_2 \to ... \to t_n]$, *where t_i is an action term for any $1 \le i \le n - 1$, and t_n is an atomic action term.*

Additionally, we assume that: $(\forall i \leq n-1)(\exists t_{i\alpha}, t_{i\beta}, t_{i\gamma}, t_{i\delta}), [(t_i = t_{i\alpha} * t_{i\beta} * t_{i\gamma}) * (t_{i+1} = t_{i\alpha} * t_{i\beta} * t_{i\delta}) * ((t_{i\gamma} \rightarrow t_{i\delta})$ is an action rule extracted from $S)]$.
For instance:

$[(b_1, v_1 \rightarrow w_1) * ([(b_2, v_2 \rightarrow w_2) * ... * [(b_p, v_p \rightarrow w_p)] \rightarrow [(b_1, v_1 \rightarrow w_1) * ... * [(b_{j1}, v_{j1} \rightarrow w_{j1}) * (b_{j2}, v_{j2} \rightarrow w_{j2}) * ... * (b_{jq}, v_{jq} \rightarrow w_{jq})] * ... * (b_p, v_p \rightarrow w_p)] \rightarrow (d, k_1 \rightarrow k_2)$

is an action path if:

$[(b_{j1}, v_{j1} \rightarrow w_{j1}) * (b_{j2}, v_{j2} \rightarrow w_{j2}) * ... * (b_{jq}, v_{jq} \rightarrow w_{jq})] \rightarrow (b_j, v_j \rightarrow w_j)$

is an action rule.

As an example for the purpose to use the action path notion, we take again the medical domain. The medical state of a patient is represented by values of attributes including results of medical tests. Action terms represent possible changes within values of some of these attributes which can be triggered by medications or consultations with doctors. At the same time, the description of a medical state of a patient based on these attributes may not be sufficient to identify any disease and the same what set of medications should be prescribed to the patient. In such cases, the doctor still may decide to prescribe some drugs with a goal to move the patient to a new medical state which hopefully will be less complex in terms of stating the medical diagnosis. The purpose of introducing the notion of an action path is to model this kind of reasoning which also describes the steps of a medical treatment.

Table 3.19 Information system S_1

X	Attribute a	Attribute b	Attribute c	Attribute d
x_1	a_1	b_1	c_1	H
x_2	a_2	b_1	c_2	H
x_3	a_2	b_2	c_2	H
x_4	a_2	b_1	c_1	H
x_5	a_2	b_3	c_2	H
x_6	a_1	b_1	c_2	A
x_7	a_1	b_2	c_2	H
x_8	a_1	b_2	c_1	L
y_1	a_1	b_1	c_2	$(A, \frac{1}{2})$
y_2	a_2	b_1	c_2	$(A, \frac{1}{2})$

Influence matrix associated with S is used to identify which candidate association action rules and which action paths are valid with respect to meta-actions and hidden correlations between classification attributes and decision attributes. Assume that $S = (X, A, V)$ is an information system with meta-actions $\{M_1, M_2, ..., M_n\}$ associated with S. Any candidate association action rule extracted from S which is covered by a subset of meta-actions in

M is called association action rule. In such a case we will also say that the validation process for a candidate association action rule was positive.

Assume now that R is the set of candidate association action rules generated by the algorithm presented in Section 2.6. The last step in the process of association action rules discovery is used to identify which rules in R are covered by some meta-actions from M. In last example, r_1 is an association action rule because the meta-action M_4 covers it. Rule r_1 is also applicable to x_1 and x_4. So, the meta-action M_4 will trigger r_1 which in turn will generate two new tuples: y_1 as the result of its application to x_1 and y_2 as the result of its application to x_4. The resulting information system is of type λ [4], [5] and it is in form of Table 3.19.

New candidate association action rules can be extracted from S_1, using algorithm similar to the one presented in [48], with only one exception - before any flexible attribute is used as a splitting one, all stable attributes have to be processed first. Next their validity is verified by meta-actions and the corresponding influence matrix associated with S_1. The influence matrix shows the correlations among classification attributes triggered off by meta-actions. If the candidate actions rules are not on par with them, then they are not classified as association action rules. However, the matrix may not show all the interactions between classification attributes, so still some of the resulting rules may fail when tested on real data. Now, if any new association action rules are extracted, then S_1 will be updated again and the process will continue till the fix point is reached (information system is not changed). The validation process for action paths with respect to meta-actions and hidden correlations between classification attributes and the decision attribute is similar to the validation process for association action rules, where verify if action rules used to define each path are valid with respect to meta-actions and correlations between classification and decision attributes.

References

1. Adomavicius, G., Tuzhilin, A.: Discovery of actionable patterns in databases: the action hierarchy approach. In: Proceedings of International Conference Knowledge Discovery and Data Mining, KDD 1997, pp. 111–114. AAAI Press (1997)
2. Agarval, R., Imielinski, T., Swami, A.: Mining associations between sets of items in massive databases. In: Proceedings of the ACM SIGMOD Conference on Management of Data, pp. 207–216 (1993)
3. Agarval, R., Srikant, R.: Fast algorithm for mining association rules. In: Proceeding of the Twentieth International Conference on VLDB, pp. 487–499 (1994)
4. Dardzinska, A.: Chase methods in incomplete information systems, PhD thesis, Polish Academy of Science, Warsaw (2002)
5. Dardzińska, A., Raś, Z.W.: Extracting Rules from Incomplete Decision Systems: System ERID. In: Lin, T.Y., Ohsuga, S., Liau, C.-J., Hu, X. (eds.) Foundations and Novel Approaches in Data Mining. SCI, vol. 9, pp. 143–154. Springer, Heidelberg (2006)
6. Dardzińska, A., Raś, Z.W.: Cooperative Discovery of Interesting Action Rules. In: Larsen, H.L., Pasi, G., Ortiz-Arroyo, D., Andreasen, T., Christiansen, H. (eds.) FQAS 2006. LNCS (LNAI), vol. 4027, pp. 489–497. Springer, Heidelberg (2006)
7. Dardzinska, A., Ras, Z.: On Rules Discovery from Incomplete Information Systems. In: Proceedings of the ICDM 2003 Workshop on Foundation and New Directions in Data Mining, pp. 31–35 (2003)
8. Dardzinska, A., Ras, Z.: Chasing unknown values in Incomplete Information Systems. In: Proceedings of ICDM 2003 Workshop on Foundations and New Directions of Data Mining, Melbourne, Florida, pp. 24–30. IEEE Computer Society (2003)
9. Dardzinska, A., Ras, Z.: Rule based Chase algorithm for partially incomplete information systems. In: Proceedings of AM 2003, Maebashi City, Japan, pp. 42–52 (October 2003)
10. Geffner, H., Wainer, J.: Modeling action, knowledge and control. In: Proceedings of the 13th European Conference on Artificial Intelligence, ECAI 1998, pp. 532–536. John Wiley & Sons (1998)

11. Gimpel, J.: A reduction technique for prime implicant tables. In: Proceedings of the Fifth Annual Symposium on Switching Circuit Theory and Logical Design, pp. 183–191 (1964)
12. Gouda, K., Zaki, M.J.: Efficiently mining maximal frequent itemsets. In: Proceedings of ICDM 2001, pp. 163–170 (2001)
13. Grzymala-Busse, J.: A new version of the rule induction system LERS. Fundamenta Informaticae 31(1), 27–39 (1997)
14. Grzymala-Busse, J.: Managing Uncertainty in Expert Systems. Kluwer Academic Publishers (1991)
15. Han, J., Pei, J., Yin, Y., Mao, R.: Mining Frequent Patterns without Candidate Generation: A Frequent-Pattern Tree Approach. Data Mining and Knowledge Discovery 8(1), 53–87 (2004)
16. He, Z., Xu, X., Deng, S., Ma, R.: Mining action rules from scratch. Expert Systems with Applications 29(3), 691–699 (2005)
17. Hodzic, Z., Bjekovic, G., Mikic, B., Radovcic, V.: Early verticalization and obesity as risk factors for development of flat feet in children. Acta Kinesiologica 2, 14–18 (2008)
18. Im, S., Raś, Z.W.: Action Rule Extraction from a Decision Table: ARED. In: An, A., Matwin, S., Raś, Z.W., Ślęzak, D. (eds.) ISMIS 2008. LNCS (LNAI), vol. 4994, pp. 160–168. Springer, Heidelberg (2008)
19. Im, S., Ras, Z., Tsay, L.-S.: Action Reducts. In: Kryszkiewicz, M., Rybinski, H., Skowron, A., Raś, Z.W. (eds.) ISMIS 2011. LNCS (LNAI), vol. 6804, pp. 62–69. Springer, Heidelberg (2011)
20. Im, S., Ras, Z., Wasyluk, H.: Action rule discovery from incomplete data. Knowl. Inf. Syst. 25(1), 21–33 (2010)
21. Kaur, H.: Actionable rules: issues and new directions, Transactions on Engineering, Computing and Technology, World Informatica Society, 61–64 (2005)
22. Komorowski, J., Skowron, A., Horn, A.: Handbook of Data Mining and Knowledge Discovery. Oxford University Press (1999)
23. Kryszkiewicz, M.: Representative Association Rules. In: Wu, X., Kotagiri, R., Korb, K.B. (eds.) PAKDD 1998. LNCS, vol. 1394, pp. 198–209. Springer, Heidelberg (1998)
24. Lin, T.Y., Yao, Y.Y., Louie, E.: Value Added Association Rules. In: Chen, M.-S., Yu, P.S., Liu, B. (eds.) PAKDD 2002. LNCS (LNAI), vol. 2336, pp. 328–333. Springer, Heidelberg (2002)
25. Lin, C.-J., Lai, K.-A., Kuan, T.-S., Chou, Y.-L.: Correlating factors and clinical significance of flexible flatfoot in preschool children. J. Pediatr. Orthop. 21, 378–382 (2001)
26. Liu, B., Hsu, W., Chen, S.: Using general impressions to analyze discovered classification rules. In: Proceedings of KDD 1997 Conference, Newport Beach, CA, pp. 31–36. AAAI Press (1997)
27. Padmanabhan, B., Tuzhilin, A.: A Belief-driven method for discovering unexpected patterns. In: Proceedings of KDD 1998, pp. 94–100 (1998)
28. Pauk, J., Ezerskiy, V., Raso, J., Rogalski, M.: Epidemiologic factors affecting plantar arch development in children with flat feet. Journal of the American Pediatric Medical Association 102(2), 114–121 (2012)
29. Pawlak, Z.: Information systems-theoretical foundations. Information Systems Journal 6, 205–218 (1981)
30. Pawlak, Z.: Rough sets - theoretical aspects of reasoning about data. Kluwer, Dordrecht (1991)

31. Penneau, K., Lutter, L.D., Winter, R.B.: Pes planus radiographic changes with foot orthoses and shoes. Foot Ankle 2, 299–303 (1982)
32. Piatetsky-Shapiro, G.: Discovery, analysis and presentation of strong rules. In: Piatetsky-Shapiro, G., Frawley, W.J. (eds.) Knowledge Discovery in Databases. AAAI (1991)
33. Quinlan, J.: Induction of decision trees. Machine Learning 1, 81–106 (1986)
34. Ras, Z.: Cooperative knowledge-based systems. Intelligent Automation and Soft Computing Journal 2(2), 193–205 (1996)
35. Ras, Z.: Collaboration control in distributed knowlegde-based systems. Information Sciences Journal 96(3/4), 193–205 (1997)
36. Ras, Z.: Dictionaries in a distributed knowledge-based system. In: Proceedings of Concurrent Engineering: Research and Applications Conference, Pittsburgh, August 29-31, pp. 383–390. Concurrent Technologies Corporation (1994)
37. Ras, Z.: Fault-recovery and intelligent distributed systems. In: Proceedings of the 20th ISMVL in Charlotte, NC, USA, pp. 372–377. IEEE Computer Society (1990)
38. Ras, Z.: Query answering based on distributed knowledge mining. In: Zhong, N., Liu, J., Ohsuga, S., Bradshaw, J. (eds.) Intelligent Agent Technology, Research and Development, Proceedings of IAT 2001, Maebashi City, Japan, pp. 17–27. Word Scientific Publishers (2001)
39. Ras, Z.: Reducts-driven query answering for distributed knowledge systems. International Journal of Intelligent Systems 17(2), 113–124 (2002), Peters, J.F., Skowron, A.: Special Issue on Rough Sets Approach to Knowledge Discovery
40. Raś, Z.W., Dardzińska, A.: Action Rules Discovery, a New Simplified Strategy. In: Esposito, F., Raś, Z.W., Malerba, D., Semeraro, G. (eds.) ISMIS 2006. LNCS (LNAI), vol. 4203, pp. 445–453. Springer, Heidelberg (2006)
41. Raś, Z.W., Dardzińska, A.: Action Rules Discovery Based on Tree Classifiers and Meta-actions. In: Rauch, J., Raś, Z.W., Berka, P., Elomaa, T. (eds.) ISMIS 2009. LNCS, vol. 5722, pp. 66–75. Springer, Heidelberg (2009)
42. Raś, Z.W., Dardzińska, A.: Action Rules Discovery without Pre-existing Classification Rules. In: Chan, C.-C., Grzymala-Busse, J.W., Ziarko, W.P. (eds.) RSCTC 2008. LNCS (LNAI), vol. 5306, pp. 181–190. Springer, Heidelberg (2008)
43. Ras, Z., Dardzinska, A.: From Data to Classification Rules and Actions, Rough Sets. Theory and Applications (2011)
44. Ras, Z., Dardzinska, A.: Ontology Based Distributed Autonomous Knowledge Systems. Information Systems International Journal 29(1), 47–58 (2004); International Journal of Intelligent Systems 26(6), 572–590
45. Ras, Z., Joshi, S.: Query approximate answering system for an incomplete DKBS. Fundamenta Informaticae Journal 30(3/4), 313–324 (1997)
46. Ras, Z., Joshi, S.: Query approximate answering system for an incomplete DKBS. In: Proceedings of the Workshop on Intelligent Information Systems, WIS, Dêblin, Poland, pp. 81–94. Polish Academy of Science (1996)
47. Rauch, J., Šimůnek, M.: Action Rules and the GUHA Method: Preliminary Considerations and Results. In: Rauch, J., Raś, Z.W., Berka, P., Elomaa, T. (eds.) ISMIS 2009. LNCS, vol. 5722, pp. 76–87. Springer, Heidelberg (2009)
48. Ras, Z., Dardzinska, A., Tsay, L., Wasyluk, H.: Association action rules. In: IEEE/ICDM Workshop on Mining Complex Data, pp. 83–290. IEEE Computer Society (2008)

49. Raś, Z.W., Tsay, L.-S., Dardzińska, A.: Tree-based Algorithms for Action Rules Discovery. In: Zighed, D.A., Tsumoto, S., Ras, Z.W., Hacid, H. (eds.) Mining Complex Data. SCI, vol. 165, pp. 153–163. Springer, Heidelberg (2009)
50. Ras, Z., Tzacheva, A.: Discovering semantic inconsistencies to improve action rules mining. In: Intelligent Information Systems, pp. 301–310 (2003)
51. Ras, Z., Tzacheva, A., Tsay, L.-S., Gurdal, O.: Mining for interesting action rules. In: Proceedings of IEEE/WIC/ACM International Conference on Intelligent Agent Technology, pp. 187–193 (2005)
52. Ras, Z.W., Wieczorkowska, A.A.: Action-Rules: How to Increase Profit of a Company. In: Zighed, D.A., Komorowski, J., Żytkow, J.M. (eds.) PKDD 2000. LNCS (LNAI), vol. 1910, pp. 587–592. Springer, Heidelberg (2000)
53. Raś, Z.W., Wyrzykowska, E., Wasyluk, H.: ARAS: Action Rules Discovery Based on Agglomerative Strategy. In: Raś, Z.W., Tsumoto, S., Zighed, D.A. (eds.) MCD 2007. LNCS (LNAI), vol. 4944, pp. 196–208. Springer, Heidelberg (2008)
54. Rauch, J.: Considerations on Logical Calculi for Dealing with Knowledge in Data Mining. In: Ras, Z.W., Dardzinska, A. (eds.) Advances in Data Management. SCI, vol. 223, pp. 177–199. Springer, Heidelberg (2009)
55. Saquer, J., Deogun, J.S.: Using Closed Itemsets for Discovering Representative Association Rules. In: Ohsuga, S., Raś, Z.W. (eds.) ISMIS 2000. LNCS (LNAI), vol. 1932, pp. 495–504. Springer, Heidelberg (2000)
56. Savasere, A., Omiecinski, E., Navathe, S.: An Efficient Algorithm for Mining Association Rules in Large Databases, Technical Report for Georgia Institute of Technology (1995)
57. Silberschatz, A., Tuzhilin, A.: On subjective measures of interestingness in knowledge discovery. In: Proceedings of KDD 1995 Conference, pp. 275–281. AAAI Press (1995)
58. Skowron, A.: Rough sets and boolean reasoning. Granular Computing: an Emerging Paradigm, 95–124 (2001)
59. Tan, P.-N., Steinbach, M., Kumar, V.: Introduction to Data Mining (2006)
60. Tzacheva, A.: Diversity of Summaries for Interesting Action Rule Discovery. In: Proceedings of Intelligent Information Systems, pp. 181–190. Springer (2008)
61. Tzacheva, A.A.: Summaries of Action Rules by Agglomerative Clustering. In: Ras, Z.W., Tsay, L.-S. (eds.) Advances in Intelligent Information Systems. SCI, vol. 265, pp. 259–271. Springer, Heidelberg (2010)
62. Tzacheva, A.: Algorithm for Generalization of Action Rules to Summaries: Special issue on Intelligent Information Processing and Web Mining. International Journal of Control and Cyberneticsr 39(2), 457–468 (2010)
63. Tzacheva, A., Ras, Z.: Action rules mining. International Journal of Intelligent Systems 20(6), 719–736 (2005)
64. Tzacheva, A., Raś, Z.W.: Constraint Based Action Rule Discovery with Single Classification Rules. In: An, A., Stefanowski, J., Ramanna, S., Butz, C.J., Pedrycz, W., Wang, G. (eds.) RSFDGrC 2007. LNCS (LNAI), vol. 4482, pp. 322–329. Springer, Heidelberg (2007)
65. Tzacheva, A., Ras, Z.: Association Action Rules and Action Paths Triggered by Meta-Actions. In: Proceedings of IEEE Conference on Granular Computing, Silicon Valley, CA, pp. 772–776 (2010)
66. Tsay, L.-S., Ras, Z.: Action rules discovery system DEAR, method and experiments. Journal of Experimental and Theoretical Artificial Intelligence 17(1-2), 119–128 (2005)

67. Tsay, L., Ras, Z., Dardzinska, A.: Mining E-Action Rules. In: Proceedings of 2005 IEEE ICDM Workshop in Houston, Texas, pp. 85–90 (2005)
68. Wang, K., Zhou, S., Han, J.: Profit Mining: From Patterns to Actions. In: Jensen, C.S., Jeffery, K., Pokorný, J., Šaltenis, S., Bertino, E., Böhm, K., Jarke, M. (eds.) EDBT 2002. LNCS, vol. 2287, pp. 70–87. Springer, Heidelberg (2002)
69. Wang, K., Jiang, Y., Lakshmanan, L.: Mining unexpected rules by pushing user dynamics. In: Proceedings of KDD 2003, pp. 246–255 (2003)
70. Wang, K., Jiang, Y., Tuzhilin, A.: Mining actionable patterns by role models. In: Proceedings of the 22nd International Conference on Data Engineering, pp. 16–25. IEEE Computer Society, Los Alamitos (2006)
71. Yang, Q., Cheng, H.: Mining case bases for action recommendation. In: Proceedings of the 2002 IEEE International Conference on Data Mining, p. 522. IEEE Computer Society (2002)
72. Zaki, M.J.: Scalable algorithms for association mining. IEEE Transactions on Knowledge and Data Engineering 12(3), 372–390 (2000)